Abou[t]

MW01707255

Pearson is the world's learning company, with presence across 70 countries worldwide. Our unique insights and world-class expertise comes from a long history of working closely with renowned teachers, authors and thought leaders, as a result of which, we have emerged as the preferred choice for millions of teachers and learners across the world.

We believe learning opens up opportunities, creates fulfilling careers and hence better lives. We hence collaborate with the best of minds to deliver you class-leading products, spread across the Higher Education and K12 spectrum.

Superior learning experience and improved outcomes are at the heart of everything we do. This product is the result of one such effort.

Your feedback plays a critical role in the evolution of our products and you can contact us - reachus@pearson.com. *We look forward to it.*

CLASS

7

Pearson IIT Foundation Series

Chemistry

Sixth Edition

CLASS

7

Pearson IIT Foundation Series

Chemistry

Sixth Edition

Trishna Knowledge Systems

Pearson

Photo Credits

Chapter 1 Opener: Designua. Shutterstock
Chapter 2 Opener: Inna Bigun. Shutterstock
Chapter 3 Opener: Inna Bigun. Shutterstock
Chapter 4 Opener: Margo_black. Shutterstock
Chapter 5 Opener: Maktatiana. 123rf.com
Chapter 6 Opener: Mikhail Rulkov. Shutterstock
Icons of Practice Questions: graphixmania. Shutterstock
Icons of Answer Keys: Viktor88. Shutterstock
Icons of Hints and Explanation: graphixmania. Shutterstock

Senior Editor—Acquisitions: Nandini Basu
Editor—Production: Sakshi Kansal

ISBN 978-93-528-6673-1

First Impression

Published by Pearson India Education Services Pvt. Ltd, CIN: U72200TN2005PTC057128

Head Office: 15th Floor, Tower-B, World Trade Tower, Plot No. 1, Block-C, Sector-16, Noida 201 301, Uttar Pradesh, India.
Registered Office: 4th Floor, Software Block, Elnet Software City, TS-140, Block 2 & 9, Rajiv Gandhi Salai, Taramani, Chennai 600 113, Tamil Nadu, India.
Fax: 080-30461003, Phone: 080-30461060
www.pearson.co.in, Email: companysecretary.india@pearson.com

Compositor: D Book Media
Printer in India at Sai Printo Pack Pvt. Ltd.

Brief Contents

Contents

Preface

Pearson IIT Foundation Series has evolved into a trusted resource for students who aspire to be a part of the elite undergraduate institutions of India. As a result, it has become one of the best-selling series, providing authentic and class-tested content for effective preparation—strong foundation, and better scoring.

The structure of the content is not only student-friendly but also designed in such a manner that it motivates students to go beyond the usual school curriculum, and acts as a source of higher learning to strengthen the fundamental concepts of Physics, Chemistry, and Mathematics.

The core objective of the series is to be a one-stop solution for students preparing for various competitive examinations. Irrespective of the field of study that the student may choose to take up later, it is important to understand that Mathematics and Science form the basis for most modern-day activities. Hence, utmost effort has been made to develop student interest in these basic blocks through real-life examples and application-based problems. Ultimately, the aim is to ingrain the art of problem-solving in the mind of the reader.

To ensure high level of accuracy and practicality, this series has been authored by a team of highly qualified teachers with a rich experience, and are actively involved in grooming young minds. That said, we believe that there is always scope for doing things better and hence invite you to provide us with your feedback and suggestions on how this series can be improved further.

Chapter Insights

Remember

Before beginning this chapter, you should be able to:

- define substance, mixture, mass, weight, etc.
- know the states of matter solid, liquid and gas
- basic properties of matter

Remember section will help them to memories and review the previous learning on a particular topics

Key points will help the students to identify the essential points in a chapter

Key Ideas

After completing this chapter, you should be able to:

- define matter, molecule, atom etc.
- know the arrangement of molecules in three different states of matter.

INTRODUCTION

The planet earth has endowed us with the basic necessities of life like air, water, food etc. Among these, air is the most vital for life and hence indispensable.

Layers of Atmosphere

Our atmosphere is divided into five layers on the basis of variations in chemical composition, density, temperature at different heights from the surface of the earth.

(i) Troposphere

(a) The layer of atmosphere which begins at the earth's surface and extends up to 20 km height is called **troposphere**.

Text: concepts are explained in a well structured and lucid manner

Note boxes are some add-on information of related topics

Note: Diluted H_2SO_4 is not used for the preparation of CO_2 because the calcium sulphate formed being insoluble in water forms a layer over marble chips and thus marble chips do not come in contact with dil H_2SO_4 further.

EXAMPLE

Give the electronic configuration and geometrical representation of the following elements.

(a) Flourine **(b)** Sodium **(c)** Chlorine **(d)** Carbon

SOLUTION

(a) Florine Z = 9

Electronic configuration = 2, 7

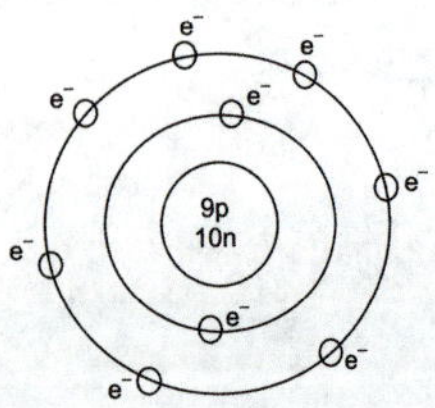

(b) Sodium Z = 11

Electronic configuration = 2, 8, 1

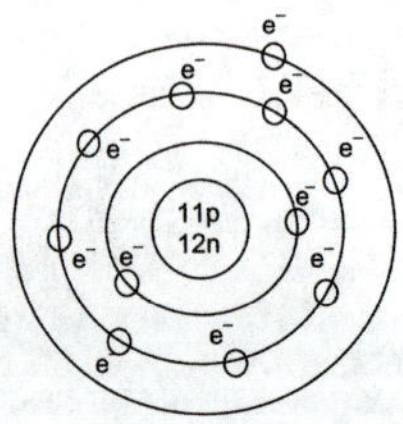

Examples given topic-wise to apply the concepts learned in a particular chapter

Illustrative examples solved in a logical and step-wise manner

TEST YOUR CONCEPTS

Very Short Answer Type Questions

Directions for questions 1 to 10: Fill in the blanks.

1. In a ______ properties of the constituents are retained.
2. A mixture of alcohol and water is an example of ______ mixture.
3. A substance which is formed by the chemical combination of two or more elements is called a ______.
4. Boiling point is the temperature at which ______ is converted to ______ at one atmospheric pressure.
5. Nonmetals usually exist in ______ state.
6. ______ show the properties of metals and nonmetals.
7. With an increase in the surface area of a liquid, the rate of evaporation______.

16. Which among the following is an element?
    (a) Calcium oxide
    (b) Common salt
    (c) Ozone
    (d) Water
17. Evaporation is the process of conversion of
    (a) a liquid to its gaseous state below the boiling point of the substance
    (b) a liquid to its gaseous state at the boiling point of the substance
    (c) a solid to its liquid state at the melting point of the substance
    (d) a solid to its liquid state below the melting point of the substance

Different levels of questions have been included in the *Test Your Concept* as well as on *Concept Application* which will help students to develop the problem-solving skills

'Test Your Concepts' at the end of the chapter for classroom preparations

CONCEPT APPLICATION

Level 1

True or false

1. Gases have maximum intermolecular spaces.
2. Metals are highly ductile and malleable.
3. Evaporation takes place from the surface of a liquid and hence it is a surface phenomenon.
4. Pure substances are made up of identical molecules and hence are homogeneous in nature.
5. Tellurium is a metalloid and hence it shows properties of both metals and nonmetals.
6. Iodine is a lustrous nonmetal.
7. As the density of sodium is less than that of water it floats on water.
8. Ammonium chloride undergoes sublimation and hence it is a sublimable substance.
9. The constituents of a compound do not retain their properties and hence these cannot be separated by physical methods.
10. Sublimation is the process of the conversion of a solid to its gaseous state.

'Concept Application' section with problems divided as per complexity: Level 1; Level 2; and Level 3

38. Carbon fibre, a recently developed allotrope of carbon is ductile and has high tensile strength.
39. Sand is insoluble in water; hence, it can be separated by filtration. Glucone-D is soluble in water; hence, it can be separated by distillation and water is collected separately.
40. Rate of evaporation is directly proportional to surface area of the liquid. This is because more number of surface molecules are exposed to atmosphere which gain kinetic energy and escape into atmosphere.
41. (i) A mixture of sawdust and water is taken in a beaker
    (ii) A filter paper is folded in the form of a cone and fitted into a funnel by moistening it with a few drops of water
    (iii) The mixture is poured gently into the filter cone and collected into another beaker which is called a filtrate.
    (iv) The solid retained on the filter paper is called a residue.

Hints and Explanation for key questions along with highlights on the common mistakes that students usually make in the examinations

Series Chapter Flow

Class 7

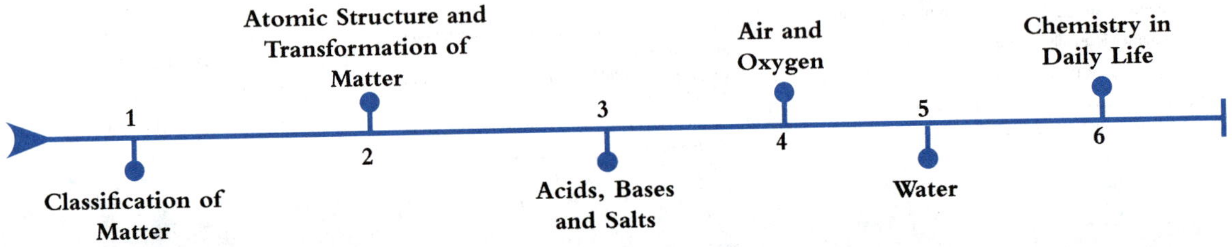

Class 9

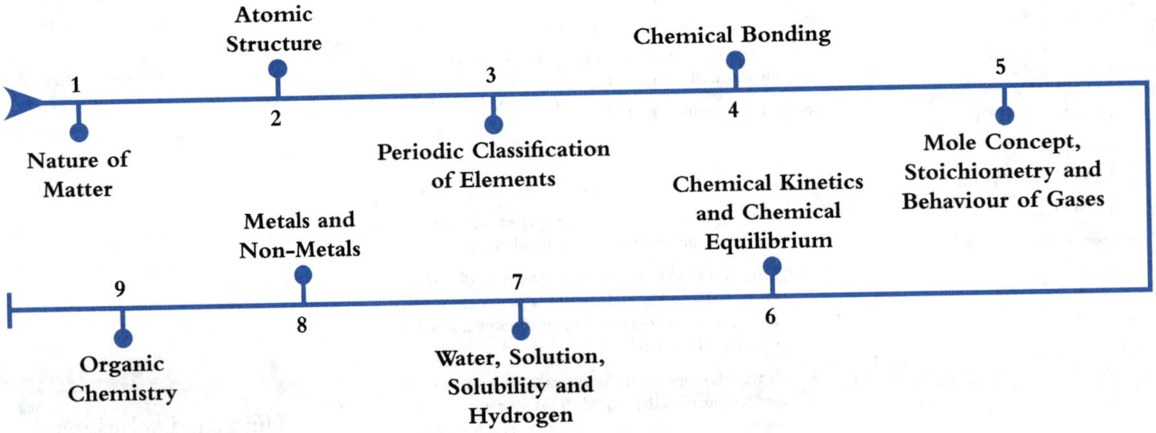

Class 8

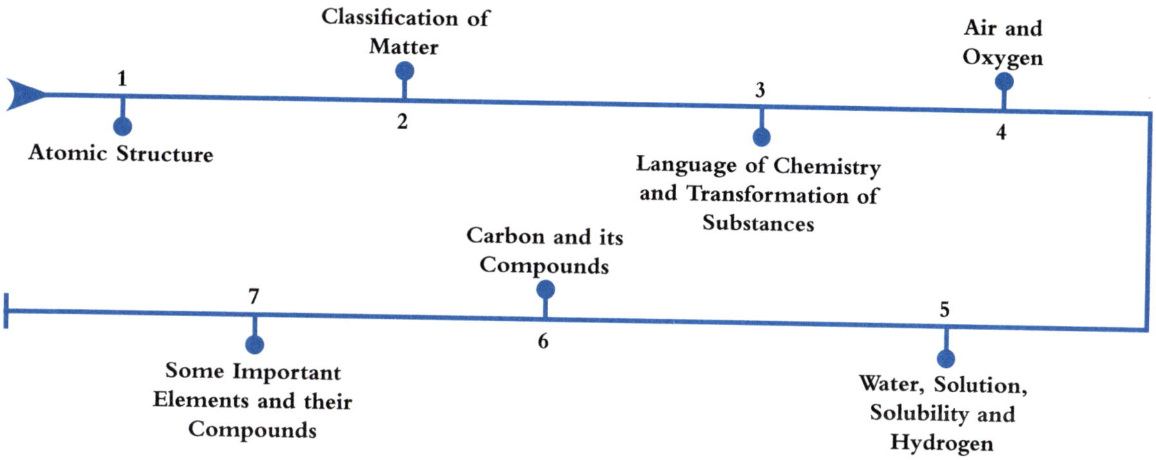

Class 10

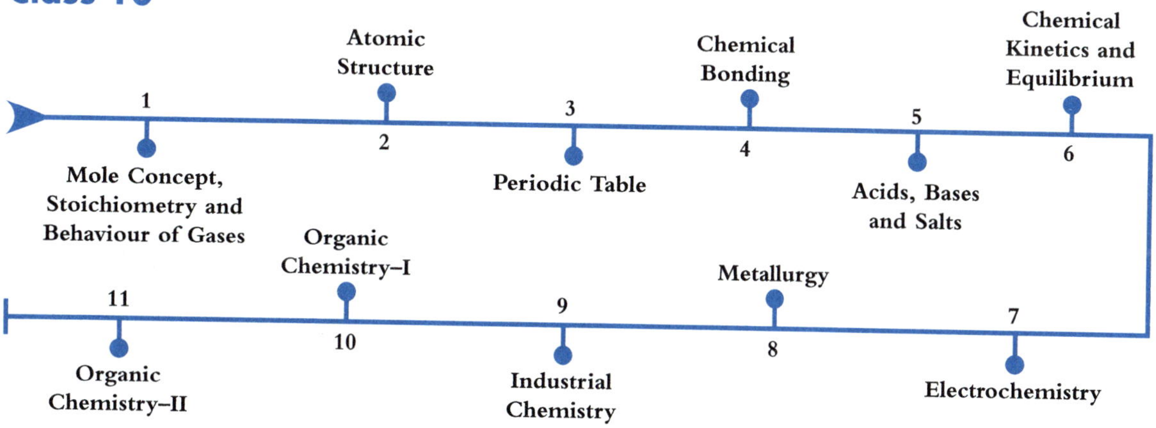

Chapter 1

Classification of Matter

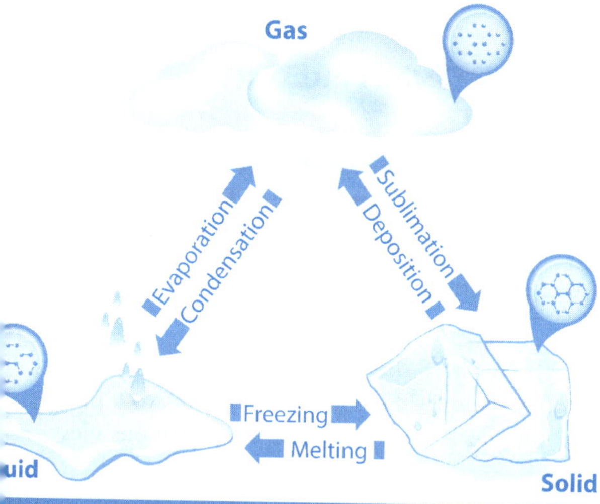

Remember

Before beginning this chapter, you should be able to:

- define substance, mixture, mass, weight, etc.
- know the states of matter—solid, liquid and gas.
- understand basic properties of matter.

Key Ideas

After completing this chapter, you should be able to:

- define matter, molecule, atom, etc.
- know the arrangement of molecules in three different states of matter.
- understand the importance of interconversion of matter and correlate it with day-to-day activities.
- learn about different kinds of matter like elements, compounds and mixtures, and also compare metals and non-metals.
- understand the different methods to separate the mixtures.

INTRODUCTION

The planet earth is endowed with a large variety of substances like rocks, water, air, coal, petroleum, etc. The beauty of the earth is attributed to the difference in physical appearance and behaviour of these substances. However, all these substances, irrespective of their physical appearance and behaviour, have certain things in common, i.e., mass and space they occupy, and are commonly called **matter**. Hence, matter can be defined as anything that occupies space and has **mass**. The basic building block of matter is known as a **molecule**. The molecule is the smallest constituent particle of matter which retains all the properties of the matter and has independent existence. The arrangement of these molecules gives rise to the variation in the physical appearance and behaviour of matter. Based on the physical appearance and behaviour, matter can be classified into three different states—**solid, liquid** and **gas**.

ARRANGEMENT OF MOLECULES IN THREE DIFFERENT STATES OF MATTER

Molecules are packed in different ways in three different states of matter which leads to the variation in other parameters like intermolecular space, intermolecular force of attraction and the movement of molecules.

Intermolecular space: The space between the constituent molecules of matter is called intermolecular space.

Intermolecular forces of attraction: The attractive forces that exist between the adjacent molecules of matter are called intermolecular forces of attraction. These forces can be of two types, i.e., cohesive forces and adhesive forces depending on the nature of the participant molecules involved.

Cohesive and Adhesive Forces

The forces of attraction between the molecules of the same substance are known as **cohesive forces**, whereas **adhesive forces** are those which exist between the molecules of two different substances. For example, water molecules stick to the glass rod due to adhesive force.

The comparative study of three different states of matter with respect to the arrangement of molecules and other related parameters is presented in Table 1.1.

TABLE 1.1 Comparative study of three states of matter

Parameter	Solids	Liquids	Gases
Packing of molecules	Very closely packed	Loosely packed	Very loosely packed
Intermolecular space	Negligible	More than that among the solid molecules	Maximum among the three states of matter
Intermolecular force of attraction	Maximum among the three states of matter	Less than that among the solid molecules	Negligible
Type(s) of movement of molecules	Vibrate to and fro about a fixed position	Vibrate to and fro, rotate around their own axis and move from one place to other within a certain limit	Vibrate to and fro, rotate around their own axis and move from one place to other randomly in all directions with much higher speed than the liquid molecules

Due to the above differences, matter around us exhibits differences in the physical properties. Table 1.2 represents the comparative study of physical properties of the different states of matter.

TABLE 1.2 Comparative study of physical properties of solids, liquids and gases

Parameter	Solids	Liquids	Gases
Mass	Definite mass	Definite mass	Definite mass
Volume	Definite volume	Definite volume	No definite volume
Shape	Definite shape	No definite shape, assume the shape of the container	No definite shape
Density	High density	Lesser density than solids	Least density
Compressibility	Incompressible	Slightly compressible	Highly compressible
Rigidity	Rigid (cannot flow)	Fluid (can flow)	Fluid (can flow)
Free surfaces	Any number of free surfaces	One free surface, i.e., only the upper surface	No free surfaces
Thermal expansion	Very low	Higher than solids	Much greater than both solids and liquids
Diffusion	Do not diffuse	Some liquids can diffuse spontaneously into another (e.g., water and alcohol), but others cannot (e.g., oil and water)	Gases diffuse spontaneously and rapidly

Interconversion of States of Matter

The existence of matter in a particular state depends on two factors, i.e., temperature and pressure. Matter can be transformed from one state to another by altering the temperature and pressure and this phenomenon is called **interconversion of states of matter**. Different types of interconversion of states of matter are given below:

1. **Melting:** The process which involves the change in state of matter from solid to liquid by heating is called **melting** or **fusion**. The temperature at which a solid changes into a liquid on heating at normal atmospheric pressure is called the **melting point** of that solid.

   Example: Ice changes to water at 0°C under normal atmospheric pressure. Hence, 0°C is the melting point of ice.

2. **Boiling:** The process which involves the change in state of matter from liquid to gas from the bulk of liquid by heating is called **boiling**. The temperature at which a liquid changes into gas from the bulk of the liquid by heating at normal atmospheric pressure is called **boiling point** of that liquid.

   Example: Water changes to water vapour at 100°C under normal atmospheric pressure. Hence, 100°C is called boiling point of water.

Every liquid exists in its vapour state over its surface at normal temperature and pressure. This is the result of change of state from liquid to vapour, i.e., evaporation. **Evaporation** can be defined as the change of state of liquid to vapour from the surface of liquid at any temperature below its boiling point. Both evaporation and boiling involve the conversion of the same state of matter, but these two phenomena can be differentiated based on the way they take place (Table 1.3).

TABLE 1.3 Comparative study of evaporation and boiling

Evaporation	Boiling
It is a slow process	It is a fast process
It takes place from the surface of a liquid	It takes place from the bulk of a liquid
The temperature of the remaining liquid decreases during evaporation	The temperature of the liquid that is undergoing boiling remains constant
It can take place at any temperature below the boiling point of the liquid	Every liquid has a fixed boiling point at a particular pressure

Factors Affecting the Rate of Evaporation

Surface area: The rate of evaporation of a liquid is directly proportional to its surface area.

Temperature: With the increase of temperature, the rate of evaporation of a liquid increases. **Humidity**: The rate of evaporation of a liquid is inversely proportional to the amount of moisture (humidity) present in the atmosphere.

3. **Condensation:** The process which involves the change of matter from gas to liquid on cooling is called **condensation.**

   Example: Conversion of water vapour or steam to water.

4. **Solidification:** The process involving the change of matter from liquid to solid on cooling is called **solidification or freezing**. The temperature at which freezing occurs at normal atmospheric pressure is called **freezing point** of that liquid.

   Example: Water changes to ice at 0°C under normal atmospheric pressure. Hence, 0°C is the freezing point of water.

   Note: Generally, solids expand on melting and liquids contract on freezing.

5. **Sublimation:** The process by which some solid substances directly change into the vapour state without passing through the intermediate liquid state is called **sublimation**. The solid obtained on cooling the vapour is called **sublimate** and the gas solid vapour formed is called **sublime**.

   Examples: Camphor, iodine, etc., get converted into their respective vapour states directly under normal atmospheric pressure.

   Matter can also be classified based on its composition in the following way irrespective of the state in which it exists.

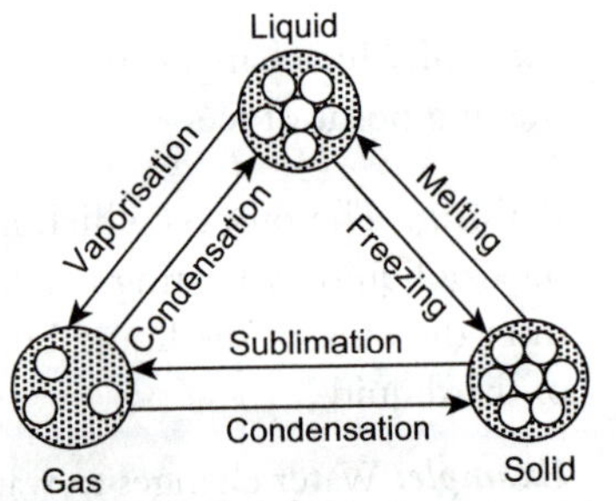

FIGURE 1.1 Schematic representation of interconversion of matter

If the matter is made up of identical molecules, it is called **pure substance**. The matter which is made up of different types of molecules is called **mixture**. Molecules of matter are made up of still tinier particles called **atoms.** Atoms are

EXAMPLE

Earthen pitchers are more effective in Hyderabad than in Chennai. Justify.

SOLUTION

The rate of evaporation of water depends on humidity (moisture present in atmosphere). The more the humidity, the less is the rate of evaporation. In coastal areas like Chennai, moisture is more, and hence, the rate of evaporation of water is less. In case of non-coastal areas like Hyderabad, moisture is less, and hence, the rate of evaporation is more. Hence, a higher rate of evaporation allows faster cooling of water in the earthen pitcher in Hyderabad when compared to Chennai.

EXAMPLE

People are advised to wear cotton clothes in summer. Give reasons.

SOLUTION

Cotton has the property of absorbing sweat. When this is exposed to atmosphere, sweat undergoes evaporation which causes cooling, and hence, we feel cool by wearing cotton clothes in summer.

the tiniest particles of matter which cannot exist independently. There are very few atoms which exist independently. These are the atoms of helium, neon, argon, krypton and xenon. They remain in the gaseous state in the atmosphere and are known as **noble gases**.

Pure substances can further be classified into **elements** and **compounds** based on the nature of constituent atoms present in their molecule.

If the molecules of a pure substance contain identical atom(s), then it is called an **element**. The pure substance in which the molecules are made up of two or more different types of atoms is called **compound**.

SCHEMATIC REPRESENTATION OF MATTER BASED ON ITS COMPOSITION

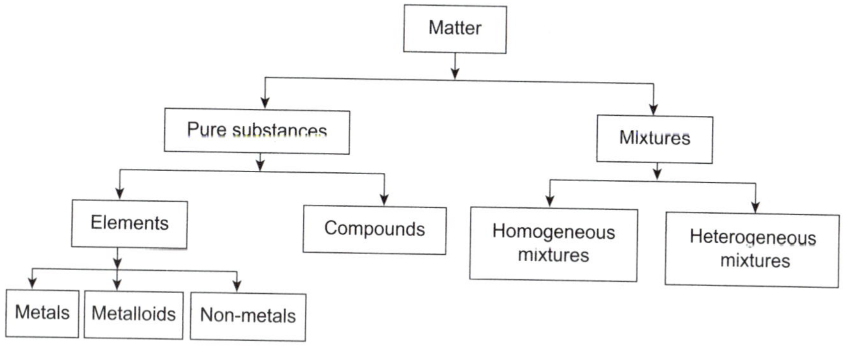

FIGURE 1.2 Classification of matter

Element

An **element** is a substance which is constituted of only one kind of atoms and cannot be further divided by any physical or chemical means.

Examples: Hydrogen, oxygen, chlorine, sodium, potassium, etc. Molecules of these substances consist of only one type of atoms. Molecules of different elements may contain different number of atoms.

Atomicity of Element

The number of atoms that constitute a molecule of an element is called **atomicity** of that element. A molecule of hydrogen is made up of two atoms, and hence, its atomicity is two. Based on the atomicity, elements can be classified as follows:

1. **Monoatomic elements**: The elements in which each molecule contains only one atom.

   Examples: Cu, Ag, He, etc.

2. **Diatomic elements**: The elements in which each molecule contains two atoms.

   Examples: H_2, O_2, N_2, etc.

3. **Polyatomic elements**: The elements in which each molecule contains more than two atoms.

   Examples: O_3, P_4, S_8, etc.

Metals and Non-metals

An element is made up of identical atoms, but the nature of the atoms is different in different elements which invariably results in differences in their characteristics. Based on the variation in properties, elements are further classified into metals, non-metals and metalloids.

TABLE 1.4 Comparative study of metals and non-metals

Property	Metals	Non-metals
State	Solids at room temperature **Exceptions**: Mercury and gallium are liquids	Mostly gases **Exceptions**: Some of the solid non-metals are carbon, sulphur, phosphorus, iodine, whereas bromine is a liquid non-metal
Melting and boiling points	Very high melting and boiling points **Exceptions**: Sodium, potassium, mercury have low melting and boiling points. Gallium has a low melting point but a high boiling point.	Low melting and boiling points. **Exceptions**: Carbon, silicon and boron have high melting and boiling points
Hardness	Generally hard **Exceptions**: Sodium and potassium are soft	Solid non-metals are brittle **Exception**: Diamond is the hardest naturally occurring substance
Density	Have high density **Exceptions**: Lithium, sodium and potassium have density lower than water (1 g/cm^3)	Have low density **Exception**: Diamond
Conductivity	Good conductors of heat and electricity **Exceptions**: Bismuth and tungsten are poor conductors of electricity	Bad conductors of heat and electricity **Exceptions**: Graphite and gas carbon are good conductors of electricity

Property	Metals	Non-metals
Lustre	Have lustre	Do not have lustre. **Exceptions**: Iodine and graphite
Tensile strength	High tensile strength **Exception**: Zinc has very less tensile strength	Do not have tensile strength. **Exception**: Carbon fibre, a recently developed allotrope of carbon
Malleability and ductility	Generally malleable and ductile. **Exception**: Zinc is neither malleable nor ductile	Non-malleable and non-ductile **Exception**: Carbon fibres are ductile
Sonorousness	Sonorous	Non-sonorous
Occurrence	Found in a combined state. **Exception**: Only noble metals are found in the free state. Silver, gold and platinum are called noble metals	Found in a free state as well as in the combined state

Metalloids

Metalloids are elements which show properties of both metals and non-metals.

Examples: Germanium, arsenic, antimony, selenium and tellurium.

Compounds

A **compound** is a substance which is formed due to the chemical combination of two or more elements in a definite proportion by mass. The constituents of a compound do not retain their properties and they can be separated only by chemical means.

Examples: Water, carbon dioxide, hydrochloric acid, sodium chloride, calcium carbonate, etc. A molecule of carbon dioxide is constituted of a carbon atom and two oxygen atoms. Two atoms of hydrogen and one atom of oxygen combine chemically to form a molecule of water.

Mixtures

A **mixture** is a substance which is formed by mixing two or more substances (elements, compounds or both) physically in any proportion. No chemical reaction takes place during the formation of a mixture and all the constituents of the mixture retain their properties. Constituents of the mixture can easily be separated by physical means.

Examples: A solution of sugar and water is a mixture in which sugar and water both retain their properties and can be separated by a simple physical process, i.e., evaporation. Muddy water is another example of mixture of mud and water which can easily be separated by filtration.

Elements and compounds are the substances in which the composition is uniform throughout the matter, and hence, they are said to be **homogeneous**. However, the composition of a mixture, i.e., the proportion of its constituents may or may not be uniform throughout. Hence, a **mixture** may be either **homogeneous** or **heterogeneous**.

Examples: A mixture of alcohol and water is an example of a homogeneous mixture because the constituents of this mixture are uniformly distributed. A mixture of sulphur and water is an example of heterogeneous mixture because sulphur remains non-uniformly distributed in water.

Some common examples of mixtures are given below:

1. Milk is a mixture of proteins, fats, vitamins, carbohydrates and mineral salts.
2. Sea water is a mixture of large number of salts and water.
3. Smoke is a mixture of air and carbon particles.
4. Cooking gas in a mixture of gases, like propane and butane.
5. Air is a mixture of nitrogen, oxygen, carbon dioxide and water vapour.

Mixtures of two or more solids are generally heterogeneous while mixtures of two or more gases are always homogeneous. However, a mixture of two or more liquids may be homogeneous or heterogeneous depending on the nature of liquids.

The liquids which completely dissolve in each other are called miscible liquids. They always form homogeneous mixtures.

Examples: Water and alcohol.

The liquids which are insoluble in each other are called immiscible liquids. They always form heterogeneous mixtures.

Examples: Coconut oil and water.

Experiments to Show the Miscibility/Immiscibility of Two Liquids

1. 30 mL of water is taken in a beaker and 15 mL of alcohol is poured into this beaker. The mixture is stirred well with the help of a glass rod. After a few minutes, it is observed that these liquid dissolve in each other forming a homogeneous mixture. Therefore, the two type of liquids are miscible liquids.
2. 30 mL of water is taken in a beaker and 15 mL of oil is poured into it and the mixture is stirred with the help of a glass rod. After a few minutes, it is observed that oil forms the upper layer and water forms the lower layer. Hence, it can be concluded that water and oil are immiscible liquids, and hence, form a heterogeneous mixture.

Comparative Study of Compounds and Mixture

TABLE 1.5 Comparative study of compounds and mixture

Compounds	Mixture
Constituent atoms of elements combine chemically	Constituent substances are mixed physically
Constituent elements do not retain their properties in a compound	The constituents retain their individual properties in a mixture
Constituents can be separated only by chemical methods	Constituents can be separated by physical methods
These are homogeneous	These can be either homogeneous or heterogeneous

Separation of a Mixture

Unlike elements, compounds are made up of more than one type of elements and a mixture is composed of more than one type of pure substances. The constituent substances of a mixture may be more than one element or compound or both elements and compounds. However, the basic difference between compound and mixture is that the constituents of the mixture retain their properties, whereas the constituent elements of the compound completely lose their original properties. Hence, constituents of the mixture can be separated easily by means of some physical processes. There are different methods of separation of the mixture based on the physical state and properties of the constituents. Some of the common processes adopted for separation of the mixtures are given below:

1. **Separation of the mixture constituting two solids**

   (a) **One constituent is a sublimable substance**

   Example: Mixture of common salt and ammonium chloride (sublimable substance). This type of mixture can be separated by a simple physical process called **sublimation process**.

Sublimation

The mixture of sodium chloride and ammonium chloride is taken in a flat basin and a funnel is inverted over it. The outer surface of the funnel is wrapped with a water-soaked filter paper and the open end of the stem of the funnel is plugged with a piece of paper or cotton to prevent vaporisation. The basin is kept in a sand bath. The sand bath is placed over a wire mesh kept on a tripod stand and then is heated gently with the help of a Bunsen burner.

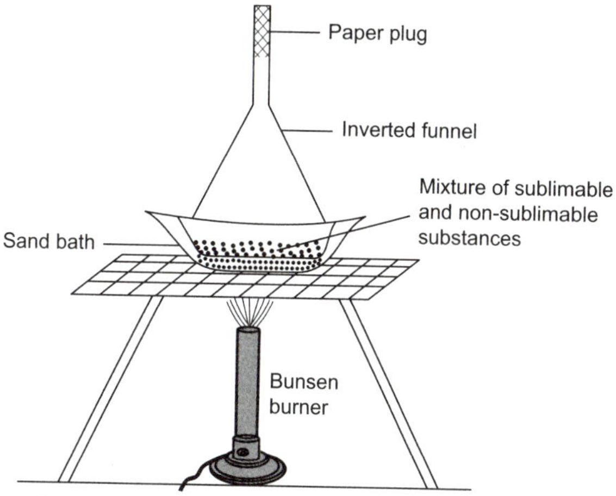

FIGURE 1.3 Separation of sublimable substance from non-sublimable substance

Observation: A white-coloured fume is observed which condenses as a white powdery substance when it comes in contact with the colder wall of the funnel. The process of heating is continued till the evolution of white vapour is observed.

Conclusion: Since ammonium chloride is a sublimable substance, therefore, it sublimes and then condenses as the white powder. The substance left over on the basin does not sublime. Hence, it is sodium chloride. Ammonium chloride is scrapped off from the wall of the funnel, and thus, the mixture is separated.

   (b) **One constituent is a magnetic substance**

   Example: A mixture of iron filings and sand can be separated by a simple physical process called magnetic separation as iron is magnetic in nature.

Process: The mixture of iron filings and sand is taken in a flat vessel. A strong bar magnet is brought close to the mixture repeatedly. The iron filings are attracted by the magnet. The iron filings that cling to the magnet can be shaved off from the surface using a piece of cardboard. Thus, iron filings are separated from the sand.

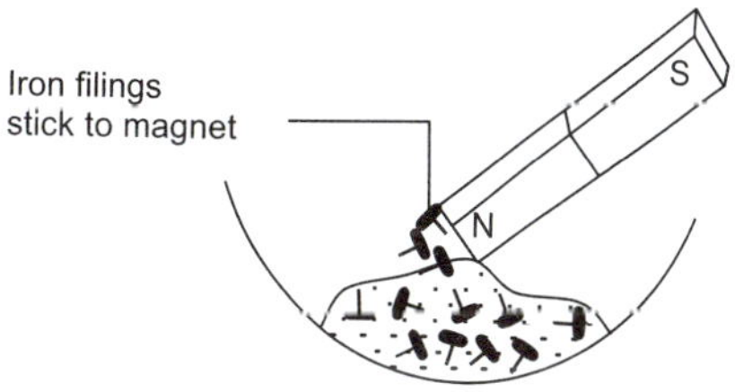

FIGURE 1.4 Separation of magnetic substance from non-magnetic substance

2. Separation of a mixture constituting one solid and one liquid

The mixture of a solid and liquid can be of different types. The solid may be soluble or insoluble in the liquid. In the case of an insoluble solid, it may float on the surface of the liquid or remain suspended in the liquid or settle down. Different methods are adopted for the separation of different types of mixture of the solid and liquid. The processes are described below:

(a) Separation of a mixture of a solid and a liquid in which the solid remains in the dissolved state

Example: Mixture of copper sulphate and water. Copper sulphate is non-volatile (the substance which does not vaporise on heating) in nature and dissolves in water. This type of mixture can be separated by **distillation process.**

Distillation Process

Process: The solution of copper sulphate is taken in a round-bottomed flask called **distillation flask**. A thermometer and a condenser, known as Liebig condenser, are attached with the round-bottomed flask with the help of adapters. Liebig condenser is a long glass tube enclosed in a jacket of glass. Cold water is circulated through the glass jacket continuously. The condenser leads to another flask called receiver to collect the liquid that gets separated from the solid during the process of distillation. The distillation flask is placed on a tripod stand and heated with the help of a Bunsen burner.

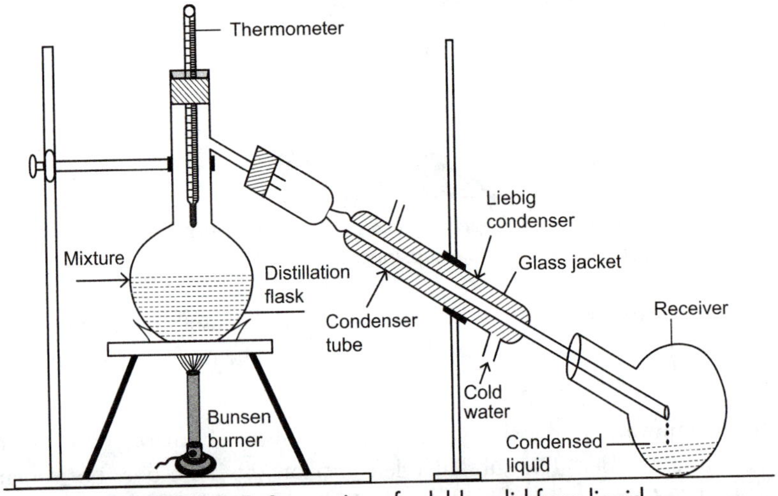

FIGURE 1.5 Separation of soluble solid from liquid

Observation: The solution starts boiling and water vapour obtained is passed through Liebig condenser where it is converted into water due to condensation and collected in the receiver. A solid residue is observed in the distillation flask.

Conclusion: Water present in the solution is vaporised leaving behind the non-volatile solid that is copper sulphate in the distillation flask, and thus, constituents of the copper sulphate solution are separated.

Evaporation is another method of separation of a solid from a solid–liquid mixture. In this method, only the solid can be obtained, but the liquid gets vaporised.

(b) Separation of a mixture of solid and liquid in which the solid does not dissolve

The insoluble solid may float or remain suspended in the liquid. If the solid particles are heavier, then they settle down.

Example: In a mixture of sawdust and water, the sawdust floats on water, fine sulphur particles remain suspended in a mixture of water and sulphur and in a mixture of sand and water, sand particles settle down if the container is kept still for some time. There are two different methods of separation, i.e., filtration and sedimentation followed by decantation for the separation of the above mixture.

Filtration

All sorts of insoluble solids can be separated from the liquid in this process.

Process: A mixture of sawdust and water is taken in a beaker. A filter paper is folded in the form of a cone and fitted into a funnel by moistening it with a few drops of water. The mixture is poured gently into the filter cone and collected into another beaker.

Observation: The liquid collected is clear and transparent. Sawdust is found over the filter paper.

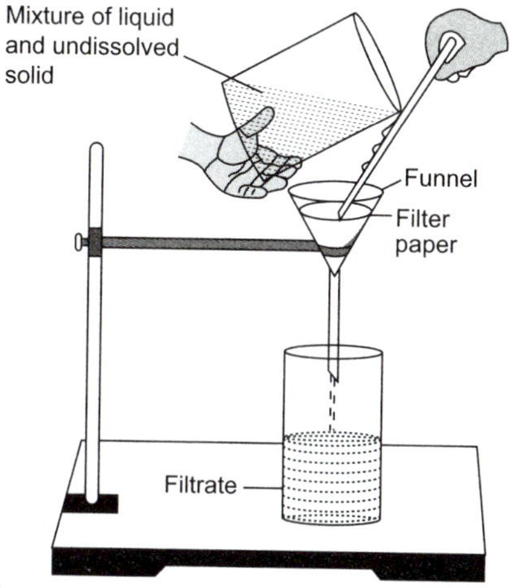

FIGURE 1.6 Separation insoluble solid from liquid

Conclusion: The mixture of sawdust and water is separated through the filtration process. The clear liquid that is collected in the other beaker is called a filtrate and the solid retained on the filter paper is known as the residue.

Sedimentation and Decantation

In the mixture of sand and water, the heavier solid particles settle at the bottom and can be separated by this process.

Process: A mixture of sand and water is taken in a beaker and allowed to stand for some time. The heavier sand particles settle down at the bottom. The clear water above it is gently poured off into another beaker with the help of a glass rod. The glass rod helps to pour off the liquid, without disturbing the solid settled at the bottom.

Observation: The water collected in the second beaker is clear and the sand particles are left behind in the first beaker.

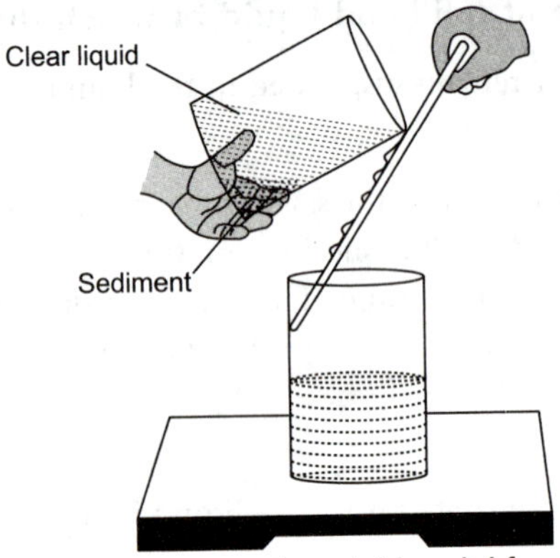

FIGURE 1.7 Separation of insoluble solid from liquid

Conclusion: Water is separated from the sand through sedimentation and decantation process.

In the above processes of separation, the mixture taken consists of only two components. In case of the mixture consisting of more than two components, separation of components from the mixture involves more than one method of separation. The methods of separation depend on the nature of constituents in the mixture.

EXAMPLE

Separation of the components of a mixture of iron filings, sand and camphor.

PROCEDURE

The mixture of iron filings, sand and camphor spread evenly in a basin. A strong bar magnet is brought close to the mixture repeatedly. The iron filings get attracted by the magnet. The iron filings that cling to the magnet can be shaved off from the surface using a piece of cardboard. The remaining constituents are covered with an inverted funnel whose outside surface is wrapped with a moist filter paper and the open end of the stem of the funnel is plugged with a piece of paper or cotton to prevent vaporisation. Then, the mixture is gently heated. Camphor undergoes sublimation and gets deposited on the cooler wall of the funnel and later on it is scrapped off. Sand is left behind in the basin. Thus, all the constituents of the given mixture are separated by magnetic separation followed by sublimation.

EXAMPLE

Separation of the components (potassium nitrate [nitre] + carbon + sulphur) of gunpowder.

PROCEDURE

Carbon disulphide is added to a beaker containing gunpowder. Sulphur gets dissolved in carbon disulphide. This mixture is then filtered with the help of a filter paper. The filtrate obtained is taken in a flat vessel and exposed to air till carbon disulphide vaporises leaving sulphur behind. The residue obtained contains carbon powder and nitre. It is then transferred to a beaker containing hot water. Nitre dissolves in hot water. The mixture is then filtered with the help of a filter paper. Carbon powder is left on the filter paper as a residue which is collected after drying. The filtrate, on evaporation to dryness by heating, gives solid nitre. Thus, the constituents of gunpowder are separated.

The processes of separation of mixtures discussed above are simple and convenient to carry out, however, complete separation of the constituents is not possible through these processes. Despite this, these simple physical processes find widespread applications in different industries and chemical laboratories.

EXAMPLE

'All pure substances are homogeneous in nature.' Justify.

SOLUTION

All pure substances contain particles of only one kind having definite set of properties and they are uniformly distributed. So, we can say that all pure substances are homogeneous in nature.

EXAMPLE

Establish logically that water is a compound.

SOLUTION

A pure substance is one which is made up of molecules containing same kind of atoms. In water, all the molecules are made up of two hydrogen atoms and one oxygen atom. These constituents are present in a fixed proportion by weight, and hence, it is a compound.

POINTS TO REMEMBER

- The intermolecular forces of attraction between similar molecules and dissimilar molecules are called cohesive and adhesive forces, respectively.
- Solids, liquids and gases possess characteristic properties based on their molecular arrangements.
- Matter exists in different states at different ranges of temperature and pressure.
- Conversion of solid to liquid is called melting and the temperature at which a solid is converted into its liquid state at atmospheric pressure is called the melting point of the solid.
- Conversion of a liquid to its gaseous state is called vaporisation which can be accomplished by two processes namely boiling and evaporation. The first one is a bulk phenomenon and the second one is a surface phenomenon.
- Rate of evaporation is directly proportional to the surface area of the liquid and temperature, and is inversely proportional to humidity.
- Evaporation causes cooling due to the absorption of heat from the surroundings.
- Based on the molecular composition, matter is classified into pure substances and mixtures. Pure substances are further classified into elements and compounds.
- The components of a mixture retain their properties, whereas the properties of a compound are totally different from those of its constituent elements.
- Components of a mixture are separated by different physical methods based on the difference in their physical properties.

TEST YOUR CONCEPTS

Very Short Answer Type Questions

Directions for questions from 1 to 10
Fill in the blanks.

1. In a ______ properties of the constituents are retained.
2. A mixture of alcohol and water is an example of ______ mixture.
3. A substance which is formed by the chemical combination of two or more elements is called a ______.
4. Boiling point is the temperature at which ______ is converted into ______ at normal atmospheric pressure.
5. Non-metals usually exist in ______ state.
6. ______ show the properties of metals and non-metals.
7. With an increase in the surface area of a liquid, the rate of evaporation______.
8. The conversion of a liquid to its solid state on cooling is called ______.
9. During freezing, heat is ______.
10. Atomicity of ozone is ______.

Directions for questions from 11 to 30
For each of the questions, four choices have been provided. Select the correct alternative.

11. Generally metals
    (a) are solids
    (b) are good conductors of heat
    (c) have high tensile strength
    (d) all the above
12. A mixture of chalk powder and ammonium chloride can be separated by ______.
    (a) distillation (b) evaporation
    (c) filtration (d) sublimation
13. Identify the soft metal among the following:
    (a) iron (b) aluminium
    (c) magnesium (d) potassium
14. Which among the following is a bad conductor of electricity?
    (a) zinc (b) copper
    (c) aluminium (d) phosphorus
15. The smallest particle of an element which may or may not have independent existence is called a/an
    (a) atom (b) molecule
    (c) compound (d) ion
16. Which among the following is an element?
    (a) Calcium oxide
    (b) Common salt
    (c) Ozone
    (d) Water
17. Evaporation is the process of conversion of
    (a) a liquid to its gaseous state below the boiling point of the substance.
    (b) a liquid to its gaseous state at the boiling point of the substance.
    (c) a solid to its liquid state at the melting point of the substance.
    (d) a solid to its liquid state below the melting point of the substance.
18. The most convenient way of separating sawdust from water is
    (a) distillation
    (b) evaporation
    (c) filtration
    (d) sedimentation and decantation
19. Which among the following statements is true?
    (a) Compounds are heterogeneous in nature.
    (b) The proportion of constituent elements in a compound is fixed.
    (c) The constituent elements retain their properties in a compound.
    (d) The formation of a compound is a physical process.
20. Which of the following is a heterogeneous mixture?
    (a) A mixture of water and sugar.
    (b) A mixture of water and common salt.
    (c) A mixture of water and sawdust.
    (d) A mixture of water and glucose.
21. Which among the following is a heterogeneous mixture?
    (a) soda water (b) liquid ammonia
    (c) milk (d) sugar water
22. Which among the following is a pure substance?
    (a) dilute sulphuric acid
    (b) concentrated sulphuric acid
    (c) aqueous NaCl
    (d) molten NaCl

23. Identify the true statement among the following:
 (a) Gases are highly compressible and diffuse very easily.
 (b) Gases are highly compressible and possess strong forces.
 (c) Solid molecules are closely packed and highly compressible.
 (d) Solid molecules are loosely packed and possess strong forces.

24. Property exploited in the usage of perfumes is_______.
 (a) compressibility of gases
 (b) diffusion of gases
 (c) expansibility of gases
 (d) both (b) and (c)

25. Baking soda ($NaHCO_3$) is a compound because
 (a) the constituents retain their properties.
 (b) the constituents can be separated by physical methods.
 (c) the constituents are combined chemically.
 (d) it is heterogeneous in nature.

26. Which among the following pairs possess low melting points?
 (a) magnesium, mercury
 (b) mercury, manganese
 (c) sodium, potassium
 (d) calcium, manganese

27. A form of which of the following non-metal is the hardest substance?
 (a) phosphorus (b) sulphur
 (c) iodine (d) carbon

28. Preparation of salt from sea water involves
 (a) evaporation
 (b) filtration
 (c) sedimentation and decantation
 (d) sublimation

29. Arrange the following in a proper sequence for the separation of constituents of gunpowder.
 (a) carbon disulphide is added to the mixture.
 (b) carbon powder is separated.
 (c) residue is filtered.
 (d) KNO_3 is separated from its aqueous solution by heating it.
 (e) mixture is filtered and hot water is added to the residue.
 (1) aedbc
 (2) aecbd
 (3) abcd
 (4) None of these

30. The steps involved in the separation of camphor and sand from a mixture are given below. Arrange them in a proper sequence.
 (a) The wet cloth is placed over the funnel and the stem is closed with a cotton plug.
 (b) The mixture of camphor and sand is taken in the china dish and an inverted funnel is kept on it.
 (c) The vapours are cooled and condensed to form the same solid and sand left behind in the dish.
 (d) The mixture is heated gently where the vapours of camphor is formed.
 (1) bcda
 (2) dabc
 (3) badc
 (4) cdbac

Directions for questions from 31 to 33

Match the entries given in Column A with the appropriate ones in Column B. More than one suitable content of Column B can also be considered correct for an entry in Column A and some content given in Column B may not be the appropriate one for any entry in Column A.

31.

Column A		Column B
A. Distillation	( )	a. One of the component goes into vapour state on heating
B. Filtration	( )	b. Separation of mixture by evaporation and subsequent condensation
C. Sedimentation	( )	c. Insoluble solids can be separated from a liquid
D. Sublimation	( )	d. In a mixture, heavier solid particles are allowed to settle and are separated from clear solution
	( )	e. A mixture of sand and iodine

PRACTICE QUESTIONS

32.

Column A		Column B
A. Atom	( )	a. Takes the shape of the container
B. Solid	( )	b. Smallest particle of the matter
C. Liquid	( )	c. Highly compressible
D. Gas	( )	d. Any number of free surfaces
E. Molecule	( )	e. Rigid
		f. Basic building block of the matter

33.

Column A		Column B
A. Boiling	( )	a. Gas to liquid
B. Melting	( )	b. Solid to gas
C. Condensation	( )	c. Liquid to solid
D. Freezing	( )	d. Liquid to gas
	( )	e. Solid to liquid

Directions for questions from 34 to 53

Answer the following questions.

34. Define cohesive and adhesive forces.

35. Define melting and boiling points.

36. Name the factors which affect the rate of evaporation.

37. Define atomicity.

38. What is the difference between an element and a compound?

39. Give two examples of metals which can exist in liquid state below 35°C.

40. What is a metalloid? Give two examples.

41. What type of metals are found in free state?

42. Distinguish sublimate from sublime.

43. What is sublimation?

44. What is meant by intermolecular space and inter-molecular forces of attraction?

45. Mention the factors on which the existence of matter in a particular state depends.

46. Distinguish between melting and boiling.

47. 'All pure substances are homogeneous in nature.' Justify.

48. How can the constituents of compounds and a mixture be separated respectively?

49. Give two examples for sublimable substances.

50. Name any two metals which are poor conductors of electricity.

51. Highlight the difference between a filtrate and a residue.

52. Name two elements which have atomicity more than three.

53. Name two states of matter which are fluid.

Short Answer Type Questions

54. Distinguish between evaporation and boiling.

55. Classify the elements based on atomicity with examples.

56. Distinguish between compounds and mixtures.

57. Explain the suitable method of separation of ammonium chloride from common salt.

58. Explain the process of sedimentation and decantation.

59. What is a homogeneous and a heterogeneous mixture? Give one example for each.

60. Explain classification of matter based on its molecular composition with suitable examples.

61. Complete the following table.

Matter	Dimension	Molecular movement
	Only one free surface	
Solid		
		Spontaneous and rapid diffusion of molecules

62. Explain the factors affecting the rate of evaporation.

63. Discuss the characteristics of
    (a) an element (b) a mixture
    (c) a compound

64. Distinguish between an atom and a molecule. Give example of elements existing as atoms and molecules.

65. Distinguish between malleability and ductility with examples.

PRACTICE QUESTIONS

66. What is intermolecular force of attraction? Mention two types of intermolecular force of attraction acting between molecules.

67. What is diffusion? Give one application.

68. Give schematic representation of classification of matter based on its composition.

69. Describe the process of separation of mixture of iron filings, camphor and sand.

Essay Type Questions

70. Compare solids, liquids and gases with respect to their physical properties.

71. Compare physical properties of metals and non-metals.

72. Explain the following methods of separation with a suitable example.
    (a) distillation (b) filtration

73. Explain different processes of interconversions of states of matter.

74. Describe the process of separation of constituents of gunpowder.

75. Give reasons for the following.
    (a) Leaves of plants appear to wilt in summer afternoons.
    (b) When a perfume is sprayed on hand, we feel cool.
    (c) Leaves of submerged aquatic plants contain wax coating.
    (d) Perspiration is greater in coastal areas than in non-coastal areas.

76. Explain sublimation with a suitable example.

77. Compare three different states of matter with respect to the arrangement of molecules and their related parameters.

For Hints and Explanations, please visit:
www.pearsoned.co.in/IITFoundationSeries

CONCEPT APPLICATION

Level 1

Directions for questions from 1 to 10
State whether the following statements are true or false.

1. Intermolecular space is maximum in gases.
2. Metals are highly ductile but non-malleable.
3. Evaporation is a surface phenomenon.
4. A pure substance is homogeneous in nature.
5. Tellurium shows the properties of both metals and non-metals.
6. Iodine is a lustrous metal.
7. Sodium floats on water.
8. Ammonium chloride is a sublimable substance.
9. The components of compound are separated by physical processes.
10. Sublimation is the process of the conversion of a solid to its liquid state.

Directions for questions from 11 to 20
Fill in the blanks.

11. _______ is a surface phenomenon, whereas boiling is a _______.
12. _______ state of matter is incompressible.
13. A pure substance is _______ in nature.
14. _______ is a lustrous non-metal.
15. Generally, the components of a compound are separated by _______.
16. _______ is the non-metal which is a good conductor of electricity.
17. _______ in humidity increases rate of evaporation.
18. A mixture of two solids is generally _______ while a mixture of any number of gases is _______.
19. Mixture of iron filings and sand can be separated by _______.
20. _______ and _______ are the suitable separation methods for mixtures of soluble solids in liquids.

Directions for questions from 21 to 41
For each of the questions, four choices have been provided. Select the correct alternative.

21. Identify the wrong statement among the following:
    (a) Molecules of solids possess only vibratory motion.
    (b) Solids are incompressible.
    (c) Solids have only one free surface.
    (d) Gases are highly diffusible.

22. The atomicity of which among the following is the maximum?
(a) helium (b) fluorine
(c) ozone (d) sulphur

23.

Column A		Column B
(A) Sublimation	( )	(a) NaCl + water
(B) Filtration	( )	(b) Iodine + sand
(C) Evaporation	( )	(c) Sawdust + water

(a) A →a B →c C →b (b) A →c B →a C →b
(c) A →c B →b C →a (d) A →b B →c C →a

24. Which of the following is a true statement regarding mixtures?
(a) They have variable composition.
(b) Mixtures are always homogeneous.
(c) Mixtures are always heterogeneous.
(d) None of the above.

25. A drop of water contains _____
(a) 2 atoms of hydrogen and 1 atom of oxygen.
(b) 1 molecule of hydrogen and 1 atom of oxygen.
(c) millions of molecules of water.
(d) 1 molecule of hydrogen and 1 molecule of oxygen.

26. Germanium is a _______
(a) gas (b) metal
(c) liquid (d) metalloid

27. The process of separation of components of muddy water is
(a) decantation (b) sublimation
(c) magnetic separation (d) none of the above

28. Lime water is a ______
(a) mixture (b) element
(c) compound (d) all the above

29. Water sticks to glass due to _______
(a) adhesive forces between water and glass.
(b) cohesive forces between water and glass.
(c) cohesive forces between water molecules.
(d) cohesive forces between glass molecules.

30. The temperature at which solid changes to liquid is called _______
(a) melting point (b) boiling point
(c) evaporation (d) condensation

31. A homogeneous mixture among the following is
(a) milk (b) muddy water
(c) smoke (d) air

32. The false statement among the following is
(a) every pure substance is homogeneous in nature.
(b) in compounds the constituents do not retain properties.
(c) the constituents of a mixture can be separated by physical method.
(d) during formation of mixtures there is a change in the molecular composition.

33. Which among the following has strong forces of attraction?
(a) hydrogen chloride (b) bromine
(c) fluorine (d) chlorine

34. Gases form homogeneous mixture due to their
(a) diffusiblity (b) high compressibility
(c) expansibility (d) low density

35. Washing soda (Na_2CO_3) is a compound because the constituents combine
(a) chemically in fixed ratio by weight.
(b) chemically in any ratio by weight.
(c) physically in fixed ratio by weight.
(d) physically in any ratio by weight.

36. Which among the following pairs of substances has strong intermolecular forces of attraction?
(a) bromine, mercury (b) gallium, bromine
(c) bromine, sodium (d) carbon, potassium

37. Among the following which is a pair of soft metals?
(a) sodium, potassium
(b) potassium, magnesium
(c) magnesium, calcium
(d) calcium, manganese

38. Identify the odd one among the following with respect to tensile strength as well as ductility.
(a) gas carbon (b) diamond
(c) graphite (d) carbon fibre

39. A mixture contains three components namely glucone-D, water and sand. These three can be collected separately by
(a) filtration and evaporation.
(b) filtration and sublimation.
(c) filtration and distillation.
(d) sedimentation and decantation.

PRACTICE QUESTIONS

40. Identify the false statement among the following.
   (a) Evaporation is a surface phenomenon and causes cooling.
   (b) Rate of evaporation is directly proportional to temperature.
   (c) Rate of evaporation is inversely proportional to surface area of a liquid.
   (d) Evaporation causes cooling and depends on humidity.

41. Separation of sawdust from water can be carried out by the following steps given below. Arrange them in a proper sequence.
   (a) The mixture is poured gently into the filter cone and collected in another beaker which is called filtrate.
   (b) A mixture of sawdust and water is taken in a beaker.
   (c) A filter paper is folded in the form of a cone and fitted into a funnel by moistening it with a few drops of water.
   (d) Solid retained on the filter paper is called residue.
   (1) cabd (2) abcd
   (3) cdba (4) bcad

Directions for questions from 42 to 44

Match the entries given in Column A with the appropriate ones in Column B. More than one suitable content of Column B can also be considered correct for an entry in Column A and some content given in Column B may not be the appropriate one for any entry in Column A.

42.

Column A		Column B
A. Sodium	( )	a. Homogeneous mixture
B. Sodium chloride	( )	b. Element
C. Sulphur in water	( )	c. Compound
D. Sugar in water	( )	d. Heterogeneous mixture
	( )	e. Impure compound

43.

Column A		Column B
A. Boiling	( )	a. Water changes to ice
B. Melting	( )	b. Water vapour changes to water
C. Condensation	( )	c. Ice changes to water
D. Freezing	( )	d. Water changes to water vapour
	( )	e. Ice changes to water vapour

44.

Column A		Column B
A. Distillation	( )	a. Mixture of sodium chloride and ammonium chloride
B. Filtration	( )	b. Mixture of sodium chloride and water
C. Sublimation	( )	c. Mixture of sawdust and water
D. Sedimentation	( )	d. Mixture of iron and sulphur
E. Magnetic separation	( )	e. Mixture of sand and water

Level 2

Directions for questions from 1 to 11

For each of the questions, four choices have been provided. Select the correct alternative.

1. Water shows concave meniscus in a narrow glass tube. This is because
   (a) adhesive force is stronger than cohesive force.
   (b) adhesive force is weaker than cohesive force.
   (c) cohesive and adhesive forces are equal.
   (d) of the absence of adhesive force.

2. Iron powder and powder of rust are taken in two containers X and Y, respectively. Dilute sulphuric acid is added to both the containers. Then
   (a) effervescence is observed in both the containers.
   (b) effervescence is observed in case of X but not Y.
   (c) effervescence is observed in case of Y but not X.
   (d) no effervescence is observed in both the cases.

3. X, Y and Z containers are placed at 25°C. Then, decrease in temperature is more in ______.

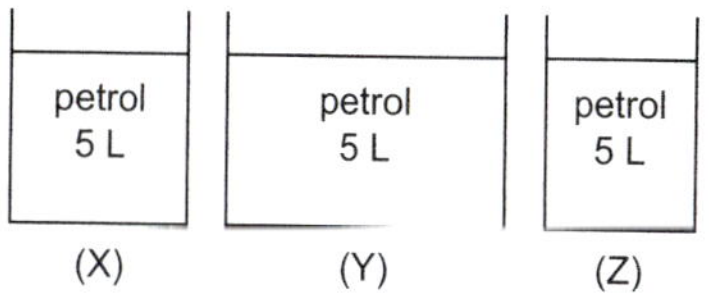

   (a) X
   (b) Y
   (c) Z
   (d) Cannot be predicted.

4. The interconversion involved in usage of 'odonil' in a washroom is______.
   (a) sublimation (b) deposition
   (c) melting (d) freezing

5. Dogs stretch out their tongues generally in summer because
   (a) evaporation leads to cooling.
   (b) of condensation of water vapour.
   (c) of freezing of saliva.
   (d) their body temperatures are high.

6. A mixture contains three components namely camphor powder, common salt and water. These can be separated by
   (a) filtration and distillation.
   (b) filtration, sedimentation and decantation.
   (c) sublimation and distillation.
   (d) sublimation, sedimentation and decantation.

7. Thermal expansion of solids is the least among the three states of matter due to
   (a) high kinetic energy of molecules of solids.
   (b) close packing of molecules in solids.
   (c) the vibratory motion and rotatory motion of molecules of solids.
   (d) the large intermolecular space present in solids.

8. Water shows convex meniscus in narrow ____ and _____ tubes.
   (a) glass, plastic
   (b) glass, wax-coated glass
   (c) wax-coated glass, plastic
   (d) plastic, coloured glass

9. Among the following, a pair of a compound and an element respectively is
   (a) iron powder and rust powder.
   (b) rust powder and iron powder.
   (c) lime and rust powder.
   (d) rust powder and lime.

10. Rate of evaporation of water
    (a) is more in coastal areas than in non-coastal areas.
    (b) is more in non-coastal areas than in coastal areas.
    (c) is the same in both coastal and non-coastal areas.
    (d) cannot be predicted.

11. Sublimation is involved in ____
    (a) incense stick and odonil.
    (b) camphor and incense stick.
    (c) perfume and odonil.
    (d) naphthalene balls and camphor.

12. Sodium catches fire easily and chlorine is a harmful gas. But sodium chloride is indispensable in our daily meal. Give reasons.

13. Sand and sawdust are mixed with water. Name the techniques that can separate sand and sawdust from water.

14. Two test tubes X and Y are filled with water and mercury, respectively. After these two liquids were poured, water drops on the inner walls of X were observed but no mercury drops in Y were seen. Give reasons.

15. Cotton clothes can be made more easily wet than synthetic clothes. Explain.

16. The thermal expansion of solids is the least among the three states of matter. Explain.

17. Camphor pellets should be preserved in air-tight containers. Give reasons.

18. During summer vacation Ravi decided to go to her grandparent's place at Delhi. Since it was summer, therefore, Ravi's mother advised him to carry only cotton clothes. Why did she say so? Give reason.

19. Why are droplets of water observed on the outer walls of a glass tumbler containing ice?

20. When Rita opened the perfume bottle in the bedroom without the permission of her mother, how did her mother came to know while watching TV in the drawing room?

21. Jack was getting late for the school, so, his mother advised him to pour hot milk from a glass into a saucer and then drink. Why did she say so? Give reason.

22. Why do we observe fog and mist in winter mornings?

23. When Vasu was running high temperature, her mother was nursing him by placing a wet cloth on his forehead till the temperature had gone down. Justify what purpose did the mother's action serve.

PRACTICE QUESTIONS

24. A student has a mixture consisting of charcoal and sulphur powder. He adds a certain reagent where he observes that one of the components goes into the solution state. Name the techniques by which the components can be separated.

25. Mother took her two daughters Bhavani and Shivani to a swimming pool in summer vacation. They both enjoyed swimming hours together. When they came out of swimming pool, both were shivering and fighting for the towel. Explain the reason for their shivering immediately after they came out of the swimming pool.

Level 3

1. How is milk powder prepared from milk?
2. Why is mercury used in a thermometer? Is it suitable for measuring high temperatures? Name the liquid which can be used for the measurement of high temperatures. Give reasons.
3. Two thermometers A and B are dipped in water and alcohol, respectively taken in two containers of similar dimensions at room temperature. Compare the temperatures shown by these two thermometers giving appropriate reasons.
4. Naturally occurring diamonds are sometimes found in different colours. Give reasons.

ASSESSMENT TEST

Test 1

Directions for questions from 1 to 12

Select the correct alternative from the given choices.

1. Identify the true statement among the following:
   (a) The constituents of both a compound and a mixture can be separated by physical methods only.
   (b) The constituents of both a compound and a mixture can be separated by chemical methods only.
   (c) The constituents of a compound and a mixture can be separated by chemical and physical methods, respectively.
   (d) The constituents of a compound and a mixture can be separated by physical and chemical methods, respectively.

2.

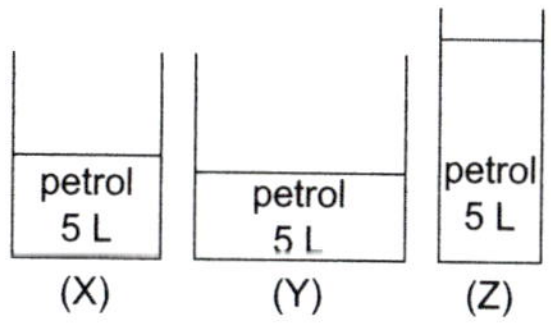

If X, Y and Z containers are placed at 25°C, then the rate of evaluation is more in __________.
   (a) X   (b) Y
   (c) Z   (d) Cannot be predicted.

3. Which among the following substances has the strongest intermolecular forces of attraction?
   (a) steam   (b) bromine
   (c) oxygen   (d) hydrogen chloride gas

4. Thermal expansion of solids is the least among the three states of matter due to
   (a) high kinetic energy of molecules of solids.
   (b) close packing of molecules in solids.
   (c) vibratory and rotatory motions of molecules of solids.
   (d) large intermolecular space present in solids.

5. **Assertion (A)**: Dogs stretch out their tongues in summer.
   Reason (R): Evaporation leads to cooling.
   (a) Both A and R are true and R is the correct explanation for A.
   (b) Both A and R are true but R is not the correct explanation for A.
   (c) A is true and R is false.
   (d) A is false and R is true.

6. Which among the following is a pure substance?
   (a) dilute sulphuric acid
   (b) concentrated sulphuric acid

(c) aqueous NaCl

(d) molten NaCl

7. Arrange the following examples in the order of matter, compound, element and mixture.

(a) dilute acid (b) argon

(c) water (d) ball

(1) a, c, b, d (2) d, c, b, a

(3) a, b, d, c (4) d, c, a, b

8. Which among the following is a heterogeneous mixture?

(a) soda water (b) air

(c) milk (d) sugar water

9. For the separation of components of a mixture of camphor, iron filings and sand, arrange the following processes in a sequence.

(a) magnetic separation

(b) distillation

(c) sublimation

(d) sedimentation and decantation

(1) a, b, c, d (2) b, a

(3) b, d (4) a, c

10. Match the following.

Column A	Column B
Principle	Procedure
(A) Evaporation ( )	(a) Purification of drinking water which contains suspended matter
(B) Filtration ( )	(b) Earthen pots
(C) Sublimation ( )	(c) Odonil used in a washroom

(a) A →a, B →c, C →b (b) A →c, B →a, C →b

(c) A →c, B →b, C →a (d) A →b, B →a, C →c

11. Property responsible for spreading of fragrance of a flower is _______.

(a) compressibility of gas/vapour

(b) diffusion of gas/vapour

(c) expansibility of gas/vapour

(d) both (b) and (c)

12. **Assertion (A)**: Baking soda ($NaHCO_3$) is a compound.

Reason (R): Properties of $NaHCO_3$ are absolutely different from sodium, carbon, hydrogen and oxygen.

(a) Both A and R are true and R is the correct explanation for A.

(b) Both A and R are true but R is not the correct explanation for A.

(c) A is true and R is false.

(d) A is false and R is true.

Test 2

Directions for questions from 1 to 9

Select the correct alternative from the given choices.

1. Identify the true statement among the following.

(a) Both compounds and mixtures are homogeneous.

(b) Both compounds and mixtures are heterogeneous.

(c) Compounds are homogeneous and mixtures are heterogeneous.

(d) Compounds are homogeneous and mixtures can be homogeneous or heterogeneous.

2. Identify the true statement among the following:

(a) Gases are highly compressible and diffuse very easily.

(b) Gases possess strong intermolecular forces of attraction.

(c) Solids are highly compressible.

(d) Solid molecules are loosely packed.

3. **Assertion (A)**: Rate of evaporation is less in rainy season.

Reason (R): Rate of evaporation is directly proportional to humidity.

(a) Both A and R are true and R is the correct explanation for A.

(b) Both A and R are true but R is not the correct explanation for A.

(c) A is true and R is false.

(d) A is false and R is true.

4. Identify the false statement among the following:

(a) A compound is homogeneous in nature.

(b) In a compound, constituents do not retain their properties.

(c) The constituents of a mixture can be separated by physical method.

(d) During formation of mixtures there is a change in the molecular composition.

5. Arrange the following substances in the ascending order of the number of constituent element(s) present in them.

(a) sodium bicarbonate (b) calcium carbonate

(c) water (d) copper

(1) d, c, b, a (2) c, b, a, d

(3) b, a, d, c (4) a, b, c, d

6. For the separation of the components of a mixture of iodine, iron filings and sawdust arrange the processes in a sequential order.

(a) The mixture is covered with an inverted funnel. The outside surface of the funnel is wrapped with a moist filter paper and the mixture is gently heated. Iodine is separated.

(b) A strong bar magnet is moved through the mixture. The iron filings are separated.

(c) Sawdust is left after iodine separates.

(d) The mixture is exposed to wind to remove sawdust.

(1) b, a, c (2) d, b, a

(3) d, c, a (4) a, d, c

7. Among the following, a pair of a compound and an element respectively is

(a) iron powder and rust powder.

(b) rust powder and iron powder.

(c) lime and rust powder.

(d) rust powder and lime.

8. Match the following:

Column A		Column B
(A) Sublimation	( )	(a) Copper sulphate and water
(B) Filtration	( )	(b) Sawdust and water
(C) Evaporation	( )	(c) Iodine and sand
(D) Magnetic separation	( )	(d) Common salt from sea water
(E) Distillation	( )	(e) Iron and sulphur

(a) A →c, B →b, C →d, D →e, E →a

(b) A →c, B →d, C →b, D →e, E →a

(c) A →d, B →e, C →c, D →a, E →b

(d) A →b, B →a, C →d, D →e, E →c

9. **Assertion (A)**: Washing soda (Na_2CO_3) is a compound.

Reason (R): Sodium retains its property in washing soda.

(a) Both A and R are true and R is the correct explanation for A.

(b) Both A and R are true but R is not the correct explanation for A.

(c) A is true and R is false.

(d) A is false and R is true.

PRACTICE QUESTIONS

TEST YOUR CONCEPTS

Fill in the blanks

1. mixture
2. homogeneous
3. compound
4. liquid, gas
5. gaseous
6. metalloids
7. increases
8. solidification
9. released
10. three

Multiple choice questions

11. d
12. d
13. d
14. d
15. a
16. c
17. a
18. c
19. b
20. c
21. c
22. d
23. a
24. d
25. c
26. c
27. d
28. a
29. b
30. c

Match the following

31. A →b
B →c
C →d
D →a, e

32. A →b
B →e, d
C →a
D →c
E →f

33. A →d
B →e
C →a
D →c

CONCEPT APPLICATION

Level 1

True or false

1. True
2. False
3. True
4. True
5. True
6. False
7. True
8. True
9. False
10. False

Fill in the blanks

11. Evaporation, bulk phenomenon
12. Solid
13. homogeneous
14. Iodine
15. chemical methods
16. Graphite
17. Decrease
18. heterogeneous, homogeneous
19. magnetic separation
20. Evaporation, distillation

Multiple choice questions

21. c
22. d
23. d
24. a
25. c
26. d
27. a
28. a
29. a
30. a
31. d
32. d
33. b
34. a
35. a
36. d
37. a
38. d
39. c
40. c
41. d

ANSWER KEYS

Match the following

42. A →b
    B →c
    C →d
    D →a

43. A →d
    B →c
    C →b
    D →a

44. A →b
    B →c
    C →a
    D →e
    E →d

Level 2

Multiple choice questions

1. a	2. b	3. b	4. a	5. a	6. a
7. b	8. c	9. b	10. b	11. d	

ASSESSMENT TEST

Test 1

1. c	2. b	3. a	4. b	5. b
6. a	7. d	8. d	9. b	10. c
11. d	12. b	13. d	14. b	15. a

Test 2

1. d	2. b	3. d	4. d	5. a
6. c	7. d	8. d	9. a	10. d
11. a	12. b	13. a	14. a	15. c

CONCEPT APPLICATION

Level 1

True or false

1. Gases have the maximum intermolecular spaces.
2. Metals are highly ductile and malleable.
3. Evaporation takes place from the surface of a liquid, and hence, it is a surface phenomenon.
4. Pure substances are made up of identical molecules, and hence, are homogeneous in nature.
5. Tellurium is a metalloid, and hence, it shows properties of both metals and non-metals.
6. Iodine is a lustrous non-metal.
7. As the density of sodium is less than that of water, it floats on water.
8. Ammonium chloride undergoes sublimation, and hence, it is a sublimable substance.
9. The constituents of a compound do not retain their properties, and hence, these cannot be separated by physical methods.
10. Sublimation is the process of the conversion of a solid to its gaseous state.

Fill in the blanks

11. Evaporation is a surface phenomenon, whereas boiling is a bulk phenomenon.
12. Solid state is incompressible as the force of attraction is strong and intermolecular space is less.
13. A pure substance is homogeneous in nature.
14. Iodine is a lustrous non-metal.
15. The components of a compound are separated by chemical methods.
16. Graphite is a non-metal and is a good conductor of electricity.
17. Decrease in humidity increases the rate of evaporation.
18. A mixture of two solids is generally heterogeneous while a mixture of any number of gases is homogeneous in nature.
19. The mixture of an iron filing and sand can be separated by magnetic separation.
20. To separate soluble solids from a liquid the suitable methods are evaporation and distillation.

Multiple choice questions

21. Solids have many free surfaces because the molecules have only vibratory motion, and hence are fixed.
22. Sulphur has the highest atomicity of eight.
23. Sublimation → Iodine + sand, iodine undergoes sublimation on heating

    Filtration → Sawdust + water, sawdust floats on water

    Evaporation → NaCl + water, NaCl is soluble in water and undergoes evaporation on heating
24. A mixture has a variable composition.
25. A drop of water contains millions of molecules of water.
26. Germanium has the properties of metals as well as non-metals, hence, it is a metalloid.
27. The process is decantation as mud being heavier than water settles down at the bottom due to gravity and separates from water.
28. Lime water is a saturated solution of calcium hydroxide, hence, it is a mixture.
29. Water sticks to glass due to adhesive forces acting between water and glass.
30. The temperature at which a solid gets converted into a liquid is called melting point.
31. Air is a homogeneous mixture of different gases.
32. During the formation of a mixture the constituent substances are mixed physically, and hence, there is no change in molecular composition during the formation of mixtures.
33. Bromine is a liquid, and hence, it possesses stronger intermolecular forces of attractions than gases.

34. Gases have the maximum intermolecular spaces, and hence, they diffuse in each other, forming a homogeneous mixture.

35. In washing soda (Na_2CO_3), the constituents are combined chemically in a fixed ratio of their weights, and hence, it is a compound.

36. Solids have strong intermolecular forces of attraction, and hence, carbon and potassium possess strong forces of attraction.

37. Soft metals are sodium and potassium

38. Carbon fibre, a recently developed allotrope of carbon is ductile and has high tensile strength.

39. Sand is insoluble in water, hence, it can be separated by filtration. Glucone-D is soluble in water, hence, it can be separated by distillation and water is collected separately.

40. Rate of evaporation is directly proportional to surface area of the liquid. This is because more number of surface molecules are exposed to atmosphere which gain kinetic energy and escape into atmosphere.

41. (i) A mixture of sawdust and water is taken in a beaker.

(ii) A filter paper is folded in the form of a cone and fitted into a funnel by moistening it with a few drops of water.

(iii) The mixture is poured gently into the filter cone and collected into another beaker which is called a filtrate.

(iv) The solid retained on the filter paper is called a residue.

Match the following

42. A → b Sodium is an element as it is made up of same atoms.

B → c NaCl is a compound because it is made up of sodium and chlorine in a fixed ratio by mass.

C → d Sulphur in water is a heterogeneous mixture as the component is non-uniformly distributed.

D → a Sugar in water is a homogeneous mixture as both the components are uniformly distributed.

43. A → d In boiling process, water changes into vapour.

B → c In melting process, ice changes into water.

C → b In condensation, water vapour changes into water.

D → a In freezing process, water changes into ice.

44. A → b In distillation process, the non-volatile component sodium chloride can be separated from water.

B → c Sawdust remains suspended in the water and separated by filtration.

C → A mixture of sodium chloride and ammonium chloride can be separated by sublimation as ammonium chloride is sublimable.

D → e Sand being heavier, settles down at the bottom and can be separated by sedimentation.

E → d Iron is magnetic and sulphur is non-magnetic in nature.

Level 2

1. The force of attraction between similar molecules is called a cohesive force and that between dissimilar molecules is called an adhesive force. Adhesive forces act between water molecules and glass molecules. Due to stronger adhesive forces over cohesive forces, water shows concave meniscus in a narrow glass tube.

2. Iron is an element and rust is a compound of iron in which iron does not retain its properties. Hence, rust does not liberate hydrogen gas with dilute sulphuric acid and no effervescence is observed in container Y.

3. Petrol is a highly volatile liquid and undergoes evaporation, which causes cooling. Rate of evaporation is directly proportional to the surface area of the liquid. In the case of container 'Y,' the rate of evaporation is more, and hence, decrease in temperature is more.

4. Odonil undergoes sublimation, i.e., solid is converted into vapour state.

5. Tongues of dogs (upon exposure to atmosphere) enable the saliva to undergo evaporation which causes cooling. Hence, the body temperature decreases.

6. Since camphor is insoluble in water, therefore, it can be separated by filtration. Common salt from water can be separated by distillation.

7. In a solid, the molecules are closely packed due to strong forces of attraction, and hence, thermal expansion of solids is least among the three states of matter.

8. In the case of a wax-coated glass and plastic tube, cohesive forces are predominant over adhesive forces, and hence, water shows convex meniscus in a narrow wax-coated glass and plastic tubes.

9. Rust is a compound (hydrated ferric oxide). Iron powder is an element.

10. Rate of evaporation is inversely proportional to humidity. In non-coastal areas humidity is less when compared to coastal areas, and hence, rate of evaporation of water is more in a non-coastal area when compared to a coastal area.

11. Naphthalene balls and camphor undergo sublimation.

12. Sodium chloride is a compound. As a compound, the constituents (sodium and chlorine) do not retain their properties, and therefore, we can take NaCl in our daily meal.

13. As the density of sand is more than sawdust, it settles down and sawdust floats on water. Sand can be separated by sedimentation and decantation. Sawdust can be separated by filtration.

14. Between water and glass molecules, adhesive forces continually act which are predominant over cohesive forces, and hence, water droplets are observed on the inner walls of the glass tube X. In the case of mercury cohesive forces are more than adhesive forces, and hence, mercury droplets are not observed on the inner walls of 'Y.'

15. The force of attraction between similar molecules is called a cohesive force and that between dissimilar molecules is called an adhesive force. Between water and cotton molecules adhesive forces act which are predominant over cohesive forces. As a result cotton clothes can be easily made wet than synthetic clothes.

16. In solids, strong intermolecular forces of attraction are present between the molecules due to which they expand less on heating.

17. The process of conversion of a solid state to a vapour state directly without passing through the intermittent liquid state is called sublimation. Camphor undergoes sublimation at atmospheric pressure. As a result, camphor pellets should be preserved in air-tight containers.

18. Cotton has the property of absorbing sweat. When this is exposed to atmosphere sweat undergoes evaporation which causes cooling, and hence, we feel cool by wearing cotton clothes in summer.

19. Ice present in a glass tumbler starts melting by absorbing heat from the air around the glass. As a result, moisture present in the air gets condensed and forms tiny water droplets on the outer walls of the container.

20. Gas particles move with high speeds and the intermolecular spaces in gases are very large. These properties of gases allow a gas to diffuse easily into another. As a result, the smell of a perfume which is in vapour state reaches quickly to the other corner of the room.

21. The rate of evaporation depends on the surface area of liquids. The more is the surface area of the liquid, the more is the rate of evaporation of the liquid. This is due to increased number of surface molecules. When these surface molecules are exposed to atmosphere, they gain energy and escape into atmosphere. Hence, hot milk in the saucer gets cooled.

22. In winters, morning temperature is very low. As a result, moisture present in the atmosphere condenses and forms tiny droplets of water. These droplets appear as fog and mist.

23. Evaporation is the process of conversion of a liquid into a gas. During this process the energy required for the conversion of the liquid into the gas is taken from the surroundings, and as a result, the temperature of the surroundings decreases, i.e., evaporation causes cooling. When a wet cloth is placed on the forehead, water present in the cloth slowly undergoes evaporation, and hence, temperature of the body decreases.

24. The reagent is carbon disulphide which dissolves sulphur. Charcoal can be separated by filtration. Sulphur from the solution can be separated by evaporation.

25. Once we come out of the swimming pool in summer, water droplets present on the body undergo evaporation (slowly) which causes cooling, and as a result we feel shivering.

Level 3

1. Milk is sent through an atomiser and as a result it is converted into small droplets, consequently the surface area increases. Due to the increased surface area, rate of evaporation of water becomes more at room temperature. In this way, the entire water undergoes evaporation leaving behind fat particles which are converted into powder by adding other ingredients.

2. Mercury expands on heating and has weak forces of attraction towards glass. Hence, it is used in thermometers. As gallium possesses high boiling point it is used for the measurement of high temperatures.

3. The rate of evaporation depends upon the nature of the liquid. In case of alcohol, the surface molecules possess high kinetic energy and as a result they easily escape into atmosphere, i.e., the rate of evaporation is more. As evaporation causes cooling, decrease in temperature in case of alcohol is more. Due to this reason, a decrease in the level of mercury is observed in the case of container B having alcohol but not in the case of A having water.

4. Naturally occurring diamonds are associated with some impurities. These impurities have characteristic colours and retain these colours in diamonds, hence, naturally occurring diamonds are found to be in different colours.

ASSESSMENT TEST

Test 1

1. The constituents of a compound do not retain their properties. Hence, they can be separated by chemical methods. The constituents of a mixture retain their properties. Hence, they can be separated by physical methods.

2. Petrol is a highly volatile liquid and undergoes evaporation. The rate of evaporation is directly proportional to the surface area of the liquid. In the case of container Y the rate of evaporation is more.

3. Bromine is a liquid, and hence, it possesses stronger intermolecular forces of attractions.

4. In a solid the molecules are closely packed due to strong forces of attraction, and hence, thermal expansion of solids is the least among the three states of matter.

5. Dogs stretch out their tongues in summer as evaporation of the saliva leads to cooling.

6. NaCl is a compound. Every compound is a pure substance. Molten NaCl means NaCl in liquid state, and hence, molten NaCl is also a pure substance.

7. d, c, b, a
   (i) ball
   (ii) water
   (iii) argon
   (iv) dilute acid

8. Milk is a heterogeneous mixture in which fat particles are present in water.

9. a, c
   (i) magnetic separation
   (ii) sublimation

10. A $\rightarrow$ b Earthen pots work on the principle of evaporation.

    B $\rightarrow$ a Purification of drinking water which contains suspended matter is based on the principle of filtration.

    C $\rightarrow$ c Odonil, used in washrooms, works on the principle of sublimation.

11. Molecules of gas/vapour diffuse easily into each other.

12. The constituents of baking soda, i.e., Na, H, C and O do not retain their properties, and hence, backing soda is a compound.

Test 2

1. Compounds are homogeneous while mixtures can be either homogeneous or heterogeneous.

2. In gases intermolecular forces of attraction are negligible, and hence, they are highly compressible and diffuse very easily.

3. Rate of evaporation is inversely proportional to humidity. In rainy season, humidity is more. Hence, rate of evaporation is less.

4. During the formation of a mixture the constituent substances are mixed physically, and hence, there is no change in molecular composition during the formation of a mixture.

5. d, c, b, a
   (i) copper
   (ii) water
   (iii) calcium carbonate
   (iv) sodium bicarbonate

6. b, a, c
   (i) A strong bar magnet is moved through the mixture. The iron filings are separated.
   (ii) The mixture is covered with an inverted funnel. The outside surface of the funnel is wrapped with a moist filter paper and the mixture is gently heated. Iodine is separated.
   (iii) Sawdust is left after iodine separates.

7. Hydrated ferric oxide is rust which is a compound. Iron powder is an element.

8. A → c Iodine and sand can be separated by sublimation process as iodine is a sublimable substance.

   B → b Sawdust and water can be separated by filtration.

   C → d We can obtain common salt from sea water by evaporation.

   D → e Iron and sulphur can be separated by the process called magnetic separation as iron gets attracted to the magnet.

   E → a Copper sulphate and water can be separated by distillation.

9. Washing soda is a compound since the constituents combine chemically.

   In a compound, constituents do not retain their properties. Since sodium is the constituent of washing soda (Na_3CO_3), therefore, it does not retain its properties.

Chapter 2

Atomic Structure and Transformation of Matter

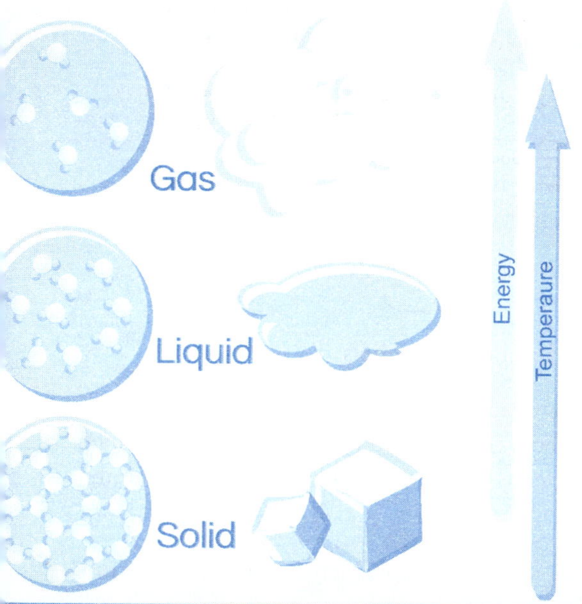

Remember

Before beginning this chapter, you should be able to:

- learn about matter and its physical forms, understand the importance of interconversion of matter and correlate it with day-to-day activities.
- learn about different kinds of matter like elements compounds and mixtures and understand the different methods to separate the mixture.

Key Ideas

After completing this chapter, you should be able to:

- learn about symbols of various elements.
- understand Dalton's atomic theory and different terms related to the structure of atom like atomic number, mass number, electronic configuration, etc.
- learn about valencies of different elements and radicals and the method of determining the formulae.
- explain the transformation of matter, types of reactions and balance chemical equations.

INTRODUCTION

From time immemorial, different scientific researches and studies have been in progress to explore the basic constituents of matter. As a result of these scientific researches and observations, it has been finally concluded that **elements** are the basic constituents of any matter, and the building block of an element is an **atom**. For the convenience of the study of matter and for the purpose of exact identification, scientists all over the world put an effort to name the element. Consequently, the names of the elements were abbreviated to **symbols** in order to represent them in an easier and a more convenient way.

Symbolic Representation of Elements

In ancient days, alchemists were under the impression that some metals like iron could be converted into gold. However, this idea remained hypothetical and could not be realised in practice.

This hypothetical art of transmutation of some suitable metals into gold was called **alchemy**.

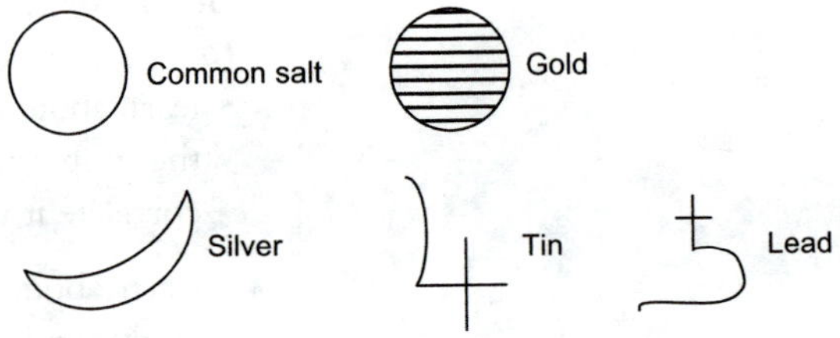

FIGURE 2.1 Examples of alchemist symbols

Symbols can be defined as the shorthand notations which represent a single atom of an element and generally consist of a letter or combination of letters derived from the name of the element. This system of writing symbols was suggested by Berzelius in the year 1814 to represent an atom of an element.

1. In most cases, the first letter of the name of the element in capital letter is used as the symbol of the element.

TABLE 2.1 Elements and their symbols as first letter of their names

Element	Symbol
Boron	B
Carbon	C
Fluorine	F
Hydrogen	H
Iodine	I
Nitrogen	N
Oxygen	O
Phosphorus	P
Sulphur	S
Vanadium	V
Uranium	U

2. When the names of two or more elements have the same initial letter (E.g., carbon and calcium), another letter from the name of the element is written in small letter along with the initial letter written in capital.

TABLE 2.2 Elements and their symbols using two letters

Element	Symbol
Aluminium	Al
Arsenic	As
Argon	Ar
Barium	Ba
Beryllium	Be
Bromine	Br
Bismuth	Bi
Calcium	Ca
Cadmium	Cd
Chromium	Cr
Chlorine	Cl
Cobalt	Co
Palladium	Pd
Platinum	Pt
Selenium	Se
Silicon	Si
Strontium	Sr

3. For some elements, the symbols are derived from their Latin, Greek or German names.

TABLE 2.3 Elements with symbols derived from their foreign names

Name of the element	Parent name	Symbol
Sodium	*Natrium* (Greek)	Na
Potassium	*Kalium* (Latin)	K
Iron	*Ferrum* (Latin)	Fe
Copper	*Cuprum* (Latin)	Cu
Silver	*Argentum* (Latin)	Ag
Gold	*Aurum* (Latin)	Au
Mercury	*Hydrargyrum* (Greek)	Hg
Lead	*Plumbum* (Latin)	Pb
Tin	*Stannum* (Latin)	Sn
Antimony	*Stibium* (Greek)	Sb
Tungsten	*Wolfram* (German)	W

The study of the atoms of different elements does not end with its nomenclature. The most important area of study of an atom is its structure. The term atom has been coined by the Greek philosophers. Indian philosopher, Maharshi Kannada named it as **Paramanu**.

In the 19th century, John Dalton, based on certain observations and experimental results, postulated a theory which regarded atoms as the fundamental building blocks of matter. This theory, based on the principle of indivisibility of atom, was well accepted throughout the century.

DALTON'S ATOMIC THEORY

In 1808, John Dalton proposed the atomic theory. Dalton's atomic theory is mainly based on the law of conservation of mass, the law of definite proportions and the law of multiple proportions which were formulated by Lavoisier.

The main postulates of Dalton's atomic theory are the following:

1. Matter is composed of tiny indivisible particles called atoms. They can neither be created nor destroyed nor transformed into atoms of another element.
2. Atoms of a given element are identical in all respects.
3. Atoms of different elements are different from each other in all respects.
4. Atoms of different elements take part in the chemical reaction and combine in a simple integral ratio to form compounds.
5. When elements react, the atoms may combine in more than one simple whole number ratio. Dalton's atomic theory has been contradicted with the advancement of science and modified on the basis of further research and discoveries.

Limitations of Dalton's Atomic Theory

1. With the discovery of subatomic particles, i.e., the electron, proton and neutron, it was concluded that atoms can be further divided.
2. Discovery of isotopes proved that atoms of the same element may possess different atomic weights, i.e., atoms of same elements may not be identical in all respects.

   Example: Isotopes of hydrogen are protium ($_1H^1$), deuterium ($_1H^2$) and tritium ($_1H^3$).
3. In some cases, atoms of different elements are found to have same mass number, i.e., atoms of different elements may be identical in this respect. For example, calcium and argon are found to have the same mass number.

Though Dalton's atomic theory could not give convincing explanation to any of the above facts, it laid down the foundation for the development of modern atomic theory.

The basic postulate of Dalton's atomic theory which says that 'atoms are the tiniest particles of matter which take part in a chemical reaction'. It is, however, accepted in modern atomic theory with experimental evidence.

However, discovery of radioactive elements led the scientists to focus on the divisibility aspect of the atom.

Later, experiments carried out by scientists like J. J. Thomson, Goldstein, Chadwick, etc., proved the presence of subatomic or fundamental particles in an atom. The changed perception that an atom is divisible made various scientists such as Rutherford, Bohr, Heisenberg, etc., to throw light on the structure of the atom precisely.

STRUCTURE OF AN ATOM

Fundamental particles: An atom consists of three types of fundamental particles, namely, **protons, electrons** and **neutrons.**

Protons: Protons are **positively charged** particles carrying a charge of $+1.6 \times 10^{-19}$ Coulombs or $+4.8 \times 10^{-10}$ esu. The charge of one proton is taken as unit positive charge (+1). The mass of one proton is 1.67×10^{-27} kg. The mass of one proton is taken as 1 atomic mass unit (amu).

Electrons: Electrons are **negatively charged** particles carrying a charge equal to that of a proton, i.e., -1.6×10^{-19} coulombs or -4.8×10^{-10} esu. The charge of one electron is taken as unit negative charge (–1). The mass of an electron is very negligible and has a value of 9.1×10^{-31} kg. The mass of an electron is $\frac{1}{1837}$ times that of the proton.

Neutrons: Neutrons do not carry any charge. The mass of a neutron is only slightly greater than that of the proton that is equal to 1.72×10^{-27} kg. The mass of one neutron is also taken as 1 amu.

Various scientists proposed various ideas regarding the arrangement of fundamental particles in an atom which is called **structure of an atom**.

The modern concept of atom envisages the following arrangement of fundamental particles in an atom:

1. The protons and neutrons are concentrated in a small region at the centre of an atom. This central part is known as **nucleus**. The protons and neutrons present inside the nucleus are called **nucleons**.
2. The size of the nucleus is very small when compared to the size of the atom. This signifies the fact that there is a vast empty space in the atom.
3. Electrons revolve round the nucleus with high speed to overcome the strong electrostatic forces of attraction by the nucleus in definite fixed paths which are called **orbits** or **shells**.
4. In an atom, the number of electrons is equal to the number of protons inside the nucleus. Since the protons and electrons carry equal and opposite charges, an atom is electrically neutral.
5. The various orbits or shells are named as **K, L, M, N**... or **1, 2, 3, 4**...... and so on. The maximum number of electrons in the various orbits is 2, 8, 18, 32......, respectively. The energy of the orbit increases with the increase in distance from the nucleus.
6. The outermost shell is called **valence shell** and it can never have more than 8 electrons in any atom.

Atomic number and **mass number**: The number of protons present inside the nucleus is called **atomic number**. It is equal to the number of electrons in the atom. The elements are characterised by their atomic numbers. The atomic number of an element is denoted by the symbol **Z**.

Example: Sodium, Z = 11, means that an atom of sodium contains 11 protons inside the nucleus and 11 electrons outside the nucleus.

Mass number: Mass number indicates the sum of protons (p) and neutrons (n) present inside the nucleus. It is denoted by the symbol **A**. The difference between the mass number and the atomic number gives the number of **neutrons**.

Mass number = Number of protons + Number of neutrons

⇒ Mass number = Atomic number + Number of neutrons

⇒A = Z + Number of neutrons

∴ Number of neutrons = A – Z

Example: Na → Z = 11, A = 23

Number of neutrons = A – Z = 23 – 11 = 12

The atomic number is the characteristic feature of each element. For the sake of convenience and universal application, all the elements are assigned unique symbols. An element with its atomic number and mass number is represented as $_ZX^A$. The arrangement of electrons in various shells around the nucleus is called **electronic configuration**. In the building up of an atom, the electrons are added to the shells in increasing order of their energy.

Filling up of electrons in the shells follows certain rules. The maximum number of electrons in an orbit is equal to $\mathbf{2n^2}$ where n denotes the number of the orbit or shell, i.e., 1, 2, 3, 4, etc. However, when the M shell is the valence shell, first it is filled with 8 electrons. The next 2 electrons go to the N shell and then the M shell gets filled with 10 more electrons because it can accommodate a maximum of 18 electrons as per the $2n^2$ formula.

Examples: **1.** In hydrogen, there is one electron which is present in the K shell and there are two electrons in the K shell of helium. Since the maximum capacity of the K shell is 2, the next electron goes into the L shell till it is filled with 8 electrons.

2. In scandium, the K, L, M and N shells are filled with 2, 8, 8, and 2 electrons, respectively and the remaining 1 electron goes to the M shell. Hence, the electronic arrangement in scandium is 2, 8, 9, and 2.

TABLE 2.4 Electronic configurations of elements up to Z = 30

Element	Atomic number	Electronic configuration			
		K	L	M	N
Hydrogen	1	1			
Helium	2	2			
Lithium	3	2	1		
Beryllium	4	2	2		
Boron	5	2	3		
Carbon	6	2	4		
Nitrogen	7	2	5		
Oxygen	8	2	6		
Fluorine	9	2	7		
Neon	10	2	8		
Sodium	11	2	8	1	
Magnesium	12	2	8	2	
Aluminium	13	2	8	3	
Silicon	14	2	8	4	
Phosphorus	15	2	8	5	
Sulphur	16	2	8	6	
Chlorine	17	2	8	7	
Argon	18	2	8	8	
Potassium	19	2	8	8	1
Calcium	20	2	8	8	2
Scandium	21	2	8	9	2
Titanium	22	2	8	10	2
Vanadium	23	2	8	11	2
Chromium	24	2	8	13	1
Manganese	25	2	8	13	2

Element	Atomic number	Electronic configuration			
		K	L	M	N
Iron	26	2	8	14	2
Cobalt	27	2	8	15	2
Nickel	28	2	8	16	2
Copper	29	2	8	18	1
Zinc	30	2	8	18	2

Geometrical Representation

The number and the arrangement of fundamental particles present in an atom can be shown diagrammatically by geometrical representation.

Examples

1. Oxygen (Z = 8)

Electronic arrangement = 2, 6

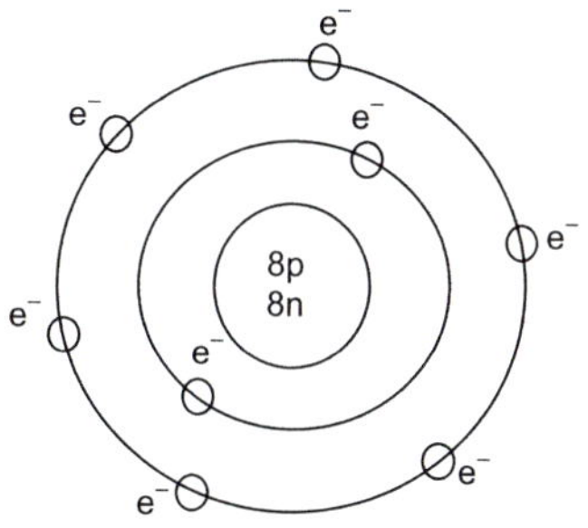

2. Phosphorus (Z = 15)

Electronic arrangement = 2, 8, 5

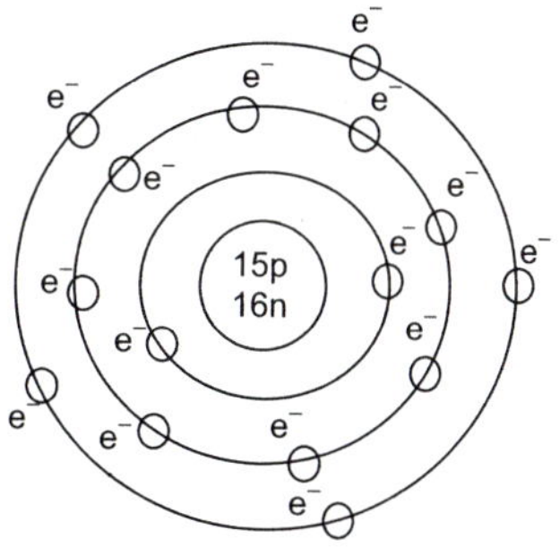

EXAMPLE

Calculate the number of neutrons in the following.

(a) $_{15}X^{31}$ **(b)** $_{92}U^{235}$ **(c)** $_{19}K^{39}$

SOLUTION

(a) Number of neutrons = 31 − 15 = 16
(b) Number of neutrons = 235 − 92 = 143
(c) Number of neutrons = 39 − 19 = 20

EXAMPLE

Give the electronic configuration and geometrical representation of the following elements.

(a) Flourine **(b)** Sodium **(c)** Chlorine **(d)** Carbon

SOLUTION

(a) Florine (Z = 9)

Electronic configuration = 2, 7

(b) Sodium (Z = 11)

Electronic configuration = 2, 8, 1

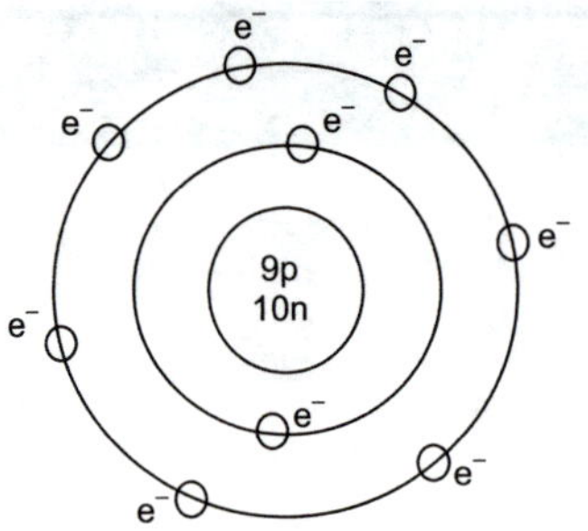

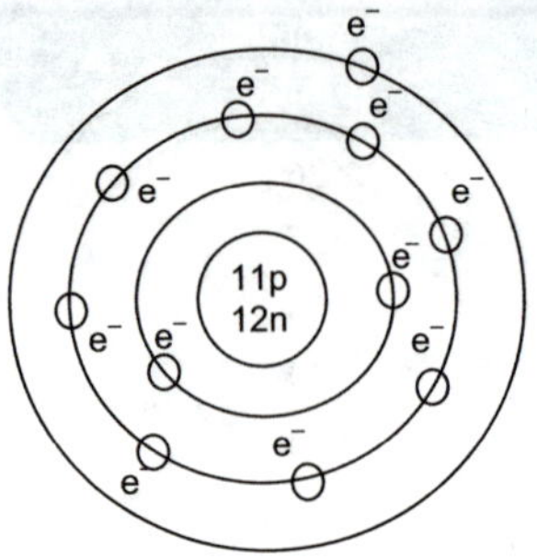

(c) Chlorine (Z = 17)

Electronic arrangement = 2, 8, 7

(d) Carbon (Z = 6)

Electronic arrangement = 2, 4

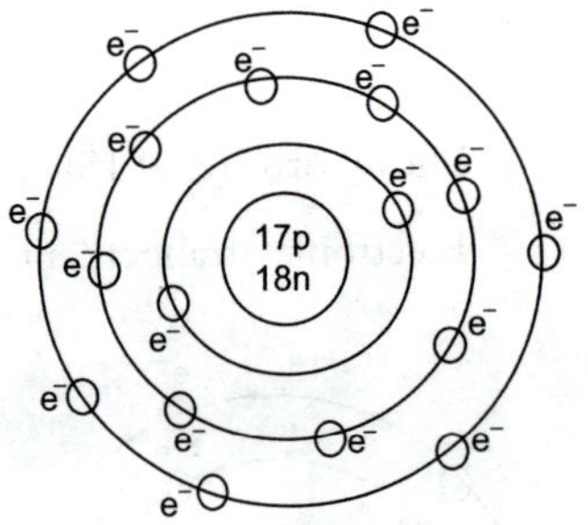

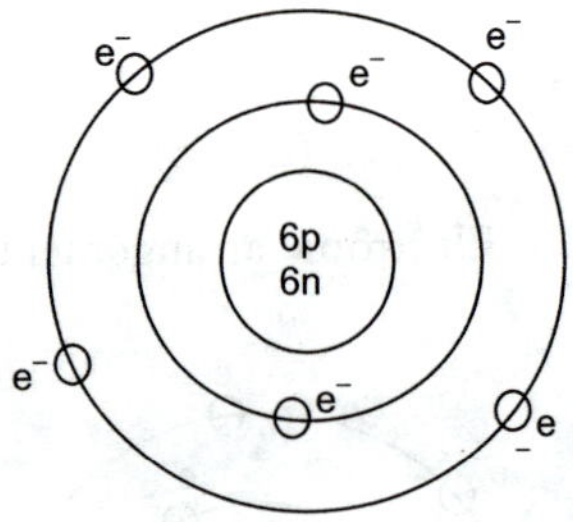

In nature, some elements are found to be quite stable as they are inactive towards chemical reactions. The atoms of these elements are found to possess eight electrons in their outermost shells. This condition is called **octet** and the resultant electronic configuration is called an octet configuration. For example, **inert gases** or **noble gases** namely neon, argon, krypton, xenon, and radon possess the octet configuration. Among the noble gases, only helium has two electrons in the valence shell and the electronic configuration is called duplet configuration. Therefore, the octet or duplet configuration is considered to be responsible for the stability of elements.

The atoms of all other elements do not possess the octet or duplet configuration. Hence, the atoms of these elements have a tendency to gain this configuration either by gaining or losing electrons.

The atoms of different elements may lose or gain different number of electrons depending on their electronic arrangement. This gives rise to the concept of valency.

Valency

The number of electrons gained or lost by an atom of an element for attaining the stable octet or duplet configuration is called **valency**.

When an atom loses one or more electrons, the number of protons becomes more than the number of electrons and the ions so formed are left with a positive charge. This is called **positive valency**.

When an atom attains stability by gaining one or more electrons, the number of electrons becomes more than the number of protons and the ion so formed becomes negatively charged. This is called **negative valency**.

The charge on the ion is corresponding to the number of electrons lost or gained. The ion may be monovalent, bivalent, trivalent and tetravalent which correspond to the valencies of 1, 2, 3, and 4, respectively.

Metals generally possess 1–3 electrons in their outermost shells and they tend to lose electrons to form positive ions called **cations**. Non-metals possess 4–7 electrons in their valence shells and they tend to gain electrons to form negative ions called **anions**.

TABLE 2.5 Valencies of different elements

Elements	Types of ion	Representation of ions	Names of ion
Sodium	Monovalent	Na^+	Sodium
Potassium	Monovalent	K^+	Potassium
Chlorine	Monovalent	Cl^-	Chloride
Fluorine	Monovalent	F^-	Fluoride
Iodine	Monovalent	I^-	Iodide
Hydrogen	Monovalent	H^-	Hydride
Hydrogen	Monovalent	H^+	Hydrogen
Zinc	Bivalent	Zn^{+2}	Zinc
Calcium	Bivalent	Ca^{+2}	Calcium
Barium	Bivalent	Ba^{+2}	Barium
Magnesium	Bivalent	Mg^{+2}	Magnesium
Oxygen	Bivalent	O^{-2}	Oxide
Oxygen	Bivalent	O_2^{-2}	Peroxide
Sulphur	Bivalent	S^{-2}	Sulphide
Aluminium	Trivalent	Al^{+3}	Aluminium
Chromium	Trivalent	Cr^{+3}	Chromium
Nitrogen	Trivalent	N^{-3}	Nitride
Phosphorus	Trivalent	P^{-3}	Phosphide

Some metals may lose different number of electrons to form ions with different charges. This is called **variable valency**.

In some cases, the electrons are also lost from the shell inner to the valence shell.

Example: iron loses two electrons from its valence shell to form Fe^{+2} and loses one more electron from the shell inner to the valence shell to form Fe^{+3}. Thus, iron exhibits variable valency.

The '**ous**' suffix is added to the name of the ion which exhibits lower valency, whereas the ion with higher valency is named by adding '**ic**' suffix.

The ions included in Tables 2.2 and 2.3 contain only one element. However, there are some charged species which contain more than one element.

TABLE 2.6 Variable valencies exhibited by some of the metals

Elements	Types of ion	Representation of ions	Names of ion
Iron	Bivalent	Fe^{+2}	Ferrous
Iron	Trivalent	Fe^{+3}	Ferric
Copper	Monovalent	Cu^+	Cuprous
Copper	Bivalent	Cu^{+2}	Cupric
Mercury	Monovalent	Hg^+	Mercurous
Mercury	Bivalent	Hg^{+2}	Mercuric
Tin	Bivalent	Sn^{+2}	Stannous
Tin	Tetravalent	Sn^{+4}	Stannic

A group of atoms possessing either positive or negative charge by losing or gaining one or more electrons is called **radical**. The positively charged radicals are ammonium radical (NH_4^+) and hydronium radical (H_3O^+). All other radicals are negatively charged.

TABLE 2.7 Some of the negative radicals

Formulas of radical	Names of radical
NO_2^-	Nitrite
NO_3^-	Nitrate
SO_3^{-2}	Sulphite
SO_4^{-2}	Sulphate
CO_3^{-2}	Carbonate
PO_4^{-3}	Phosphate
HCO_3^-	Bicarbonate or hydrogen carbonate
HSO_3^-	Bisulphite or hydrogen sulphite
HSO_4^-	Bisulphate or hydrogen sulphate
OH^-	Hydroxide
MnO_4^-	Permanganate
CrO_4^{-2}	Chromate
SiO_3^{-2}	Silicate
ZnO_2^{-2}	Zincate

In nature, a number of compounds are found. These compounds are the combination of positive and negative ions or radicals. The following example explains the formation of NaCl from its constituent atoms.

Sodium (Z = 11) has the electronic configuration of 2, 8, 1. By losing one electron, it forms Na^+ ion which attains the configuration of neon, i.e., 2, 8. Chlorine (Z = 17) has the electronic configuration of 2, 8, 7. By gaining one electron, it forms Cl^- ion which attains the configuration of argon, i.e., 2, 8, 8. Since Na^+ and Cl^- ions carry opposite charges, a strong force of attraction binds the two ions together. Thus, the compound NaCl is formed.

The compounds are represented with the help of the shorthand notations called **formulae** which are derived from the symbols of the constituent elements. Formula represents a molecule of an element or compound and is a combination of the symbols of the constituent elements along with the number of atoms of individual elements present in a molecule.

Derivation of Formulae of Compounds from the Constituent Radicals

Whatever be the valencies of the constituent radicals, a chemical compound is always electrically neutral. That means, in a molecule of a compound, the positive valencies are balanced by the negative valencies.

Example: In a molecule of aluminium chloride, aluminium ion is trivalent and carries +3 charge and chloride ions are monovalent and carry –1 charge. In order to neutralise the +3 charge on the aluminium ion, three Cl^- ions are necessary. Thus, the formula can be written as $AlCl_3$.

The formula of a compound can be derived by using a shortcut method known as **criss-cross method**.

TABLE 2.8 Examples of different compounds with their molecular composition

Names of the compound	Formulas	Molecular compositions
Hydrochloric acid	HCl	1 hydrogen atom 1 chlorine atom
Nitric acid	HNO_3	1 hydrogen atom 1 nitrogen atom
Carbon dioxide	CO_2	1 carbon atom 2 oxygen atoms
Carbon monoxide	CO	1 carbon atom 1 oxygen atom 3 oxygen atoms
Sulphur dioxide	SO_2	1 sulphur atom 2 oxygen atoms
Sulphur trioxide	SO_3	1 sulphur atom 3 oxygen atoms
Nitrous oxide	N_2O	2 nitrogen atoms 1 oxygen atom
Nitrogen dioxide	NO_2	1 nitrogen atom 2 oxygen atoms
Nitrogen trioxide	N_2O_3	2 nitrogen atoms 3 oxygen atoms
Nitrogen pentoxide	N_2O_5	2 nitrogen atoms 5 oxygen atoms
Phosphorus trioxide	P_2O_3	2 phosphorus atoms 3 oxygen atoms
Phosphorus pentoxide	P_2O_5	2 phosphorus atoms 5 oxygen atoms
Methane	CH_4	1 carbon atom 4 hydrogen atoms

1. The positive radical is written first followed by the negative radical.

   Example: Ca^{+2}, Cl^-

2. The valencies of the ions or radicals are written below the radicals without positive and negative signs.

   Example: Ca^{+2}, Cl^-
   2 1

3. The valencies should be interchanged and written as subscripts on right side of the respective ions.

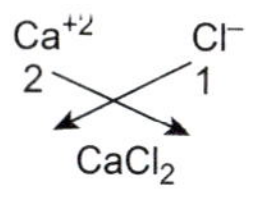

4. If the valencies have common factors, they are divided by the highest common factor.

Example:

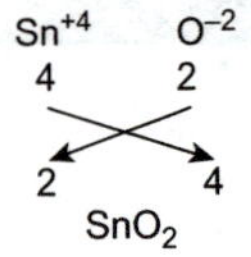

5. The radicals containing more than one atom are enclosed in brackets if the number of radicals present in a molecule is more than one.

Example:

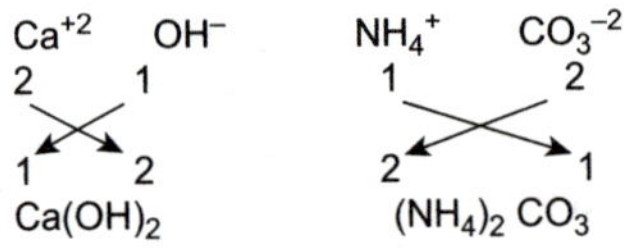

Nature is associated with a wide variety of chemical substances. Each substance has a definite molecular composition. Based on the molecular composition, the ratio of number of atoms of the constituent elements is indicated by a **chemical formula**.

Physical and Chemical Changes

The origin of the wide variety of chemical substances in different forms is attributed to the inherent tendency of matter to undergo transformation. Matter can undergo transformation in two different ways as follows:

1. Transformation that involves the alteration in certain specific properties of matter like state, colour, texture, electrical conductivity, magnetic behaviour, etc., without the alteration of molecular composition is called **physical change**.

   Example: During melting of ice, solid ice melts to liquid water. In this transformation, the physical state of ice changes, but the molecular composition remains the same. Molecules of both ice and water are composed of two hydrogen atoms and one oxygen atom. Hence, this change is called **physical change**.

2. Transformation that takes place due to the alteration of molecular composition of matter is called **chemical change**.

   Example: Magnesium ribbon burns in presence of oxygen to form white-coloured powdery magnesium oxide. Magnesium is an element and contains only magnesium atoms. Oxygen is an element which consists of only oxygen atoms. But magnesium oxide is a compound and each molecule of it is composed of one magnesium atom and one oxygen atom.

TABLE 2.9 Comparative study of physical and chemical changes

	Physical change	Chemical change
(i)	Molecular composition of the matter does not change	Molecular composition of the matter changes
(ii)	No new substance is formed. Only the physical appearance or behaviour changes	New substance is formed
(iii)	This change is called temporary or reversible because the previous form of the matter can easily be revived	This change is called permanent or irreversible because the previous form of the matter cannot be revived by physical processes

TABLE 2.10 Examples of physical and chemical changes

Physical changes	Chemical changes
(i) Melting of wax	(i) Burning of candle
(ii) Glowing of electric bulb	(ii) Electrolysis of water
(iii) Expansion of metals	(iii) Cooking of food
(iv) Changing of water into steam	(iv) Curdling of milk
(v) Boiling of milk	(v) Germination of seeds
(vi) Condensation of steam	(vi) Rusting of iron
(vii) Sublimation of camphor	(vii) Ripening of fruits
	(viii) Respiration
	(ix) Blackening of silverware
	(x) Burning of LPG

Representation of a Chemical Change in Short Form with the Help of Symbols and Formulae

The substances which undergo chemical change are called **reactants** and those formed as a result of the change are called **products**. The formulae of reactants are written on the left-hand side and the formulae of products are written on the right-hand side of an arrow. The different reactants and products are separated by a '+' sign. When a chemical change is represented in the form of an equation with the help of symbols and formulae of the reactants and the products, it is called **chemical equation**.

Example: $Mg + O_2 \rightarrow MgO$

Hydrogen and oxygen react to form water. However, a chemical equation should be written based on the principle that atoms can neither be created nor destroyed. This signifies the fact that the number of atoms of each element on either side must be equal. Such an equation is called **balanced chemical equation**.

Example: $Mg + O_2 \rightarrow MgO$ is an unbalanced chemical equation. This is also called **skeletal equation**.

A systematic procedure should be followed for balancing the chemical equation.

Step I The number of times an element occurs on both sides is counted. The element which appears the least number of times is balanced first.

Step II Among the elements with same frequency of occurrence, the metallic element is balanced first followed by non-metallic element.

Step III Hydrogen and oxygen are balanced at the end.

Examples

	R P
1. $Mg + O_2 \rightarrow MgO$	1 – Mg – 1
	2 – O – 1
$Mg + O_2 \rightarrow 2MgO$	1 – Mg – 2
	2 – O – 2
$2Mg + O_2 \rightarrow 2MgO$	2 – Mg – 2
	2 – O – 2

2. $KClO_3 \rightarrow KCl + O_2$

1 – K – 1
3 – O – 2
1 – Cl – 1

$2KClO_3 \rightarrow KCl + 3O_2$

2 – K – 1
2 – Cl – 1
6 – O – 6

$2KClO_3 \rightarrow 2KCl + 3O_2$

2 – K – 2
2 – Cl – 2
6 – O – 2

3. $Fe(OH)_3 + H_2SO_4$ (dil) $\rightarrow Fe_2(SO_4)_3 + H_2O$

1 – Fe – 2
1 – S – 3
5 – H – 2
7 – O – 13

$2Fe(OH)_3 + 3H_2SO_4 \rightarrow Fe_2 (SO_4)_3 + H_2O$

2 – Fe – 2
12 – H – 2
3 – S – 3
18 – O – 13

$2Fe(OH)_3 + 3H_2SO_4 \rightarrow Fe_2(SO_4)_3 + 6H_2O$

2 – Fe – 2
3 – S – 3
12 – H – 12
18 – O – 18

Depending on the way the products are formed from the reactants, the reactions are classified into four different types.

Types of Chemical Reactions

1. **Chemical combination:** The reaction in which two substances, i.e., elements, compounds or both combine to form a new compound is called **chemical combination**.
Representation: $X + Y \rightarrow XY$

TABLE 2.11 Types of chemical combinations with examples

Types	Examples
Element – element (synthesis)	$H_2 + Cl_2 \rightarrow 2HCl$
Element – compound	$2SO_2 + O_2 \rightarrow 2SO_3$
Compound – compound	$CaO + H_2O \rightarrow Ca(OH)_2$

2. **Chemical decomposition**: The reaction in which a compound undergoes decomposition to form more than one substance in the presence of heat or light or electricity is called chemical decomposition.

Representation: $AB \rightarrow A + B$

TABLE 2.12 Types of chemical decompositions with examples

Types	Examples
Thermal decomposition (in the presence of heat)	$2\ Pb(NO_3)_2 \rightarrow 2PbO + 4NO_2 + O_2$
Photolytic decomposition (in the presence of light)	$2HOCl \rightarrow 2HCl + O_2$
Electrolytic decomposition (by passing electricity)	$2H_2O \rightarrow 2H_2 + O_2$

3. **Chemical displacement**: When a few pieces of zinc are dipped in green-coloured ferrous sulphate solution, the colour disappears after some time and iron is found to be deposited. This phenomenon takes place because zinc is a more active metal than iron, and hence, it displaces iron from ferrous sulphate.

When copper is added to the ferrous sulphate solution, no change is observed. From this observation, it can be concluded that copper is a less active metal than iron.

The chemical reaction in which the more reactive element displaces the less reactive element from its compound is called **chemical displacement reaction.**

Representation:

$X + YZ \rightarrow XZ + Y$

X is a more reactive element than Y.

$Y + XZ \rightarrow$ No reaction

Y cannot displace X from its compound because it is less reactive than X.

Example: $Zn + FeSO_4 \rightarrow ZnSO_4 + Fe$

$Cu + FeSO_4 \rightarrow$ No reaction

Some metals liberate hydrogen gas when they are treated with acids such as HCl and dil H_2SO_4. Such metals are more active than hydrogen since they have the ability to displace hydrogen from acids. The metals which cannot displace hydrogen from acids are considered as less active metals than hydrogen.

Depending on the relative tendency of one metal to displace another metal from its compound, the metals are arranged in a series of decreasing order of reactivity. This is called **metal reactivity series.**

Metal reactivity series

Potassium	K	(Most reactive metal)
Sodium	Na	↓
Calcium	Ca	
Magnesium	Mg	
Aluminium	Al	
Zinc	Zn	Reactivity decreases
Iron	Fe	
Nickel	Ni	
Tin	Sn	
Lead	Pb	
Hydrogen	H	
Copper	Cu	
Mercury	Hg	
Silver	Ag	
Gold	Au	↓
Platinum	Pt	(Least reactive metal)

The more reactive metal displaces the less reactive metal from its compound.

4. **Chemical double decomposition/double displacement**: The chemical reaction in which the two reactants (compounds) exchange their respective radicals and two other new compounds are formed is called double displacement reaction.

Representation:

$AB + XY \rightarrow AY + XB$

Where B and Y are the negative radicals which are exchanged to give AY and XB.

Example: $2KNO_3 + FeSO_4 \rightarrow K_2SO_4 + Fe(NO_3)_2$

In the above reactive K^+ (potassium ion) combines with SO_4^- (sulphate ion) and Fe^{+2} (ferrous ion) combines with NO_3^- (nitrate ion) to form two new compounds, K_2SO_4 and $Fe(NO_3)_2$. During certain double displacement reactions, a solid substance is found to be deposited. The solid deposition is called precipitate and this type of reaction is called **precipitation reaction**.

Example: $NaCl + AgNO_3 \rightarrow NaNO_3 + AgCl \downarrow$

In the above reaction, during the exchange of radicals, silver chloride is formed which is insoluble in water. Therefore, it remains as a solid at the bottom of the container as **precipitate.**

EXAMPLE

Derive the formulae of the following compounds.

(a) Ammonium nitrate **(b)** Ferrous sulphate

SOLUTION

(a) Ammonium nitrate

$NH_4^+ \ NO_3$

∴ Formula is NH_4NO_3.

(b) Ferrous sulphate

$Fe^{+2} \ SO_4^{-2}$

∴ Formula is $FeSO_4$.

EXAMPLE

Give the necessary chemical equations and balance them.

(a) Reaction of red hot iron with steam to give Fe_3O_4.

(b) Burning magnesium ribbon in an atmosphere of nitrogen.

(c) Displacement of copper from its sulphate salt solution by zinc.

SOLUTION

(a) $3Fe + 4H_2O \longrightarrow Fe_3O_4 + 4H_2$

(Steam) Ferrosoferric oxide

(b) $3Mg + N_2 \longrightarrow Mg_3N_2$

(c) $CuSO_4 + Zn \longrightarrow ZnSO_4 + Cu$

EXAMPLE

Explain the observable changes taking place during reactions of iron with the aqueous solutions of H_2SO_4 and $CuSO_4$, respectively.

SOLUTION

Fe on treatment with H_2SO_4 liberates hydrogen gas. The solution turns green due to the formation of $FeSO_4$.

$Fe + H_2SO_4 \rightarrow FeSO_4 + H_2 \uparrow$

$Fe + CuSO_4(aq) \rightarrow FeSO_4 + Cu$

As Fe is more reactive than Cu, it displaces Cu from $CuSO_4$ solution. Hence, blue colour of $CuSO_4$ solution turns green due to the formation of $FeSO_4$ and brown-coloured solid copper is formed.

EXAMPLE

In the table given below, some compounds are listed. In each case identify the positive and negative radicals present in the compound.

Compounds	Positive radical	Negative radical
(a) SiO_2		
(b) CuS		
(c) H_2O_2		
(d) BaO_2		
(e) MgS		
(f) $NaAlO_2$		
(g) MnO_2		
(h) LiH		
(i) $KMnO_4$		
(j) $CaSiO_3$		
(k) CaF_2		
(l) HOCl		
(m) $CaCO_3$		

SOLUTION

	Compounds	Positive radical	Negative radical
(a)	SiO_2	Si^{+4}	O^{-2}
(b)	CuS	Cu^{+2}	S^{-2}
(c)	H_2O_2	H^+	O_2^{-2}
(d)	BaO_2	Ba^{+2}	O_2^{-2}
(e)	MgS	Mg^{+2}	S^{-2}
(f)	$NaAlO_2$	Na^+	AlO_2^-
(g)	MnO_2	Mn^{+4}	O^{-2}
(h)	LiH	Li^+	H^-
(i)	$KMnO_4$	K^+	MnO_4^-
(j)	$CaSiO_3$	Ca^{+2}	SiO_3^{-2}
(k)	CaF_2	Ca^{+2}	F^{-1}
(l)	$HOCl$	H^+	ClO^-
(m)	$CaCO_3$	Ca^{+2}	CO_3^{-2}

POINTS TO REMEMBER

- An atom contains a small central positive part known as nucleus.
- Nucleus contains protons and neutrons which are together called nucleons.
- Electrons revolve round the nucleus in circular paths called orbits or shells.
- The orbits are named as K, L, M, N ------- and their energies increase with the increase in distance from the nucleus.
- The outermost shell of an atom is called valence shell.
- Metals contain less number of electrons in their valence shells and non-metals contain more number of electrons in their valence shells. Therefore, they form positive and negative ions, respectively.
- A group of atoms possessing either positive or negative charge is called radical.
- The formula of a compound should be written in such a way that the positive and negative charges are neutralised.
- Physical and chemical changes can be distinguished with respect to the change in molecular composition.
- Element–element combination reactions are called synthesis reactions.
- Displacement reactions are the reactions which involve displacement of a less active element from its compound by a more active element.
- In a double decomposition reaction if one of the products formed is an insoluble solid, it is called precipitation reaction.

TEST YOUR CONCEPTS

Very Short Answer Type Questions

Directions for questions from 1 to 10
Fill in the blanks.

1. The symbol of calcium is ______.
2. Magnesium attains an octet configuration by ______ electrons.
3. The electronic arrangement of an atom with two electrons in the first shell is called ______ if the first shell is the valence shell.
4. Two molecules of sulphur trioxide are indicated by ______.
5. ______ is the suffix used while naming the salt of a metal with lower valency.
6. Name of the radical PO^{-3}_4 is ______.
7. Formulae of sodium sulphite and sodium sulphate are, respectively ______ and ______.
8. Generally ______ changes are reversible because there is no change in the ______ of the substance.
9. Decomposition of solid ammonium chloride is associated with ______ change.
10. The number of atoms of a molecule or compound is balanced on either side of a chemical equation by keeping appropriate ______.

Directions for questions from 11 to 26
For each of the questions, four choices have been provided. Select the correct alternative.

11. The number of electrons present in the K shells of hydrogen and helium are
    (a) 2 and 1, respectively
    (b) 2 in both the elements
    (c) 1 and 2, respectively
    (d) 1 in both the elements
12. The charge possessed by permanganate is the same as ______ radical.
    (a) magnesium (b) bisulphate
    (c) sulphate (d) potassium
13. If the formulae of respective chlorides of X and Y are XCl_3 and YCl_4, respectively, then the valencies of X and Y are
    (a) 3 and 2 (b) 3 and 4
    (c) 1 and 1 (d) 1 and 4
14. The formula of a compound when a positive radical with valency 2 and negative radical with valency 1 combine is ______.
    (a) B_2A (b) A_2B
    (c) AB_2 (d) AB
15. Sugar gets charred when treated with conc. H_2SO_4. This is an example of ______
    (a) physical change
    (b) chemical change
    (c) simultaneous physical and chemical changes
    (d) chemical combination reaction
16. Melting of wax is a ______.
    (a) chemical change
    (b) decomposition reaction
    (c) both physical and chemical change
    (d) physical change
17. Corrosion of a metal involves change in ______.
    (a) chemical composition
    (b) only colour and shape
    (c) only density
    (d) malleability and ductility
18. Formation of clouds, mist and fog are the examples of ______.
    (a) chemical combination of O_2 and H_2O.
    (b) physical change which involves condensation of water vapour.
    (c) physical change which involves sublimation.
    (d) chemical change which involves absorption of huge amounts of energy.
19. The reaction $AgNO_3 + NaCl \rightarrow AgCl + NaNO_3$ is a ______ reaction.
    (a) chemical displacement
    (b) chemical combination
    (c) double displacement
    (d) decomposition
20. Which of the following elements is chemically inactive?
    (a) Nitrogen (b) Neon
    (c) Hydrogen (d) Copper

21. The formula of chloride of a corresponding metal is MCl_2. What is the valency of the metal?
   (a) 4 (b) 3
   (c) 1 (d) 2

22. What will be the formulae of the sulphate and sulphite of a trivalent metal?
   (a) $M_2(SO_4)_3$, $M_2(SO_3)_3$
   (b) MSO_4, $M_2(SO_3)_3$
   (c) M_2SO_4, M_2SO_3
   (d) MSO_4, MSO_3

23. Identify the atomic numbers of the pair of elements which possesses same number of electrons in the L and M shells.
   (a) 12, 18 (b) 19, 20
   (c) 18, 16 (d) 17, 18

24. Which of the following is a physical change?
   (a) Dissolution of lime in water.
   (b) Dissolution of CO_2 in water at high pressure.
   (c) Dissolution of concentrated acid in water.
   (d) Electrolysis of water.

25. Which of the following is not a chemical change?
   (a) Passing of steam over red hot coke.
   (b) Colour change of dil nitric acid upon long standing.
   (c) Absorption of moisture by P_2O_5.
   (d) Absorption of moisture by $CaCl_2$.

26. Arrange the following points in a sequence for derivation of formulae of compounds.
   (a) The valencies of the ions or radicals are written below the radicals without positive and negative sign.
   (b) If the valencies have common factors, then they are divided by the highest common factor.
   (c) The magnitude of valencies should be interchanged and written as subscripts on the right side of the respective ions.
   (d) The positive radical followed by negative radical are written with their charges.
   (e) If the radicals contain more than one atom, they are enclosed in brackets and the appropriate valency (more than one) written as subscripts outside the bracket.
   (a) bcdea (b) dacbe
   (c) abcde (d) dabce

***Directions for questions from 27 to 29:* Match the entries given in Column A with the appropriate ones in Column B. More than one suitable content of Column B can also be considered correct for an entry in Column A and some content given in Column B may not be the appropriate one for any entry in Column A.**

27.

Column A			Column B
A.	$3O_2$	( )	a. 4 atoms of helium
B.	2Na	( )	b. 3 oxygen atoms
C.	4He	( )	c. 2 ions of sodium
D.	$8Cl_2$	( )	d. 2 atoms of sodium
		( )	e. 3 molecules of oxygen
		( )	f. 8 chlorine atoms
		( )	g. 8 molecules of chlorine

28.

Column A Name of radical			Column B Charge
A.	Zincate	( )	a. −3
B.	Stannic	( )	b. +1
C.	Nitride	( )	c. +2
D.	Potassium	( )	d. +4
		( )	e. −2

29.

Column A			Column B
A.	$X + Y \rightarrow XY$	( )	a. Displacement reaction
B.	$X + YZ \rightarrow XZ + Y$	( )	b. Combination reaction
C.	$AB + XY \rightarrow AY + XB$	( )	c. Decomposition reaction
D.	$AB \rightarrow A + B$	( )	d. Double displacement reaction

Directions for questions from 30 to 49
Answer the following questions.

30. What is the difference between symbol and formula?

31. Compare the mass of proton, neutron and electron?

32. How does the energy of orbits in an atom vary?

33. The atomic number and mass number of an element are 13 and 27, respectively. Find the number of neutrons.

PRACTICE QUESTIONS

34. Define valency?

35. Name two metals that can exhibit variable valency. Also give the names of radicals.

36. What is meant by synthesis reaction? Give an example.

37. A + BC → AC + B, In this equation, compare the reactivities of A and B.

38. What is meant by electrolytic decomposition? Give an example.

39. Give reason for the stability of inert gases.

40. What is a valence shell? What is the maximum number of electrons it can have in an atom?

41. An atom is electrically neutral. Justify.

42. What is photolytic decomposition? Give an example.

43. Identify physical and chemical changes among the following:

(a) Boiling of milk.

(b) Ripening of fruits.

(c) Curdling of milk.

(d) Melting of a candle.

(e) Rusting of iron.

(f) Blackening of silverware.

44. Why should a chemical equation be balanced?

45. Dissolution of calcium oxide in water is which type of reaction?

46. In a double decomposition reaction, one of the products is insoluble. What type of reaction is that?

47. Potassium chlorate on strong heating gives potassium chloride and oxygen. What type of reaction is this?

48. Balance the following chemical equations:

(i) $NH_3 + Br_2 \rightarrow N_2 + NH_4Br$

(ii) $FeS_2 + O_2 \rightarrow Fe_2O_3 + SO_2$

49. In the following table, some compounds are listed. In each case, identify the positive and negative radicals present in the compound.

No.	Compounds	Positive radical	Negative radical
(i)	$CaSiO_3$		
(ii)	H_2SO_3		

Short Answer Type Questions

50. Give the characteristics (symbol, charge and mass) of all fundamental particles.

51. Why do metals form cations and non-metals form anions?

52. Calculate the number of neutrons in the following.

(a) ${}_{15}X^{31}$ (b) ${}_{92}U^{235}$ (c) ${}_{19}K^{39}$

53. For each of the following reactions identify the products formed and balance the reaction.

(a) $NaCl + AgNO_3 \rightarrow$ (b) $Zn + HCl \rightarrow$

54. Define radical. Give some examples of bivalent negative radicals.

55. How do you distinguish a physical change from a chemical change?

56. Mention different types of combination reactions with examples.

57. What is meant by thermal decomposition? Give an example.

58. Define double decomposition reaction. Give an example.

59. What is a chemical displacement reaction? Mention the condition required for it.

60. Distinguish between cations and anions.

61. In the following table, some commonly used positive and negative radicals are listed. Use the criss-cross method to obtain the formulae of the compound that is formed using the given radicals. Name the compound thus obtained.

No.	Positive radicals	Negative radicals	Formula	Name of the compound
1.	Al^{+3}	PO_4^{-3}		
2.	Ca^{+2}	NO_3^{-1}		
3.	NH_4^{+1}	SO_4^{-2}		
4.	Al^{+3}	C^{-4}		

62. Distinguish between atomic number and mass number with an example.

63. Draw the geometrical representation of sulphur.

64. Derive the formulae of the following compounds.
   (a) Ammonium nitrate
   (b) Ferrous sulphate

65. How do you distinguish metals and non-metals with respect to electronic arrangement?

66. Write the electronic configurations of the following elements:
   (a) magnesium (b) silicon
   (c) sulphur (d) argon

67. Explain the formation of sodium chloride.

68. Write the symbols of the following elements.
   (a) cadmium (b) bismuth
   (c) silicon (d) argon

69. Explain the observable change taking place during reaction of iron with the aqueous solutions of H_2SO_4 and $CuSO_4$, respectively.

Essay Type Questions

70. Explain in detail the modern structure of atom.

71. Give the electronic configuration and geometrical representation of the following elements:
   (a) fluorine (b) sodium
   (c) chlorine (d) carbon

72. Classify the chemical reactions into various types depending on the way the products are formed by giving suitable examples.

73. Mention the postulates and limitations of Dalton's atomic theory.

74. Write the steps in balancing a chemical equation with an example.

75. What is valency? Write short note on different types of valencies of cations with examples.

For Hints and Explanations, please visit:
www.pearsoned.co.in/IITFoundationSeries

CONCEPT APPLICATION

Level 1

Directions for questions from 1 to 10
State whether the following statements are true or false.

1. The number of electrons in second shell of a metal with three valence electrons in the third shell is eight.
2. The valency of helium is zero.
3. The name of CrO_4^{-2} radical is dichromate ion.
4. Rusting of iron is an example of both physical and chemical change.
5. Sublimation of camphor is a physical change.
6. The central part of the atom called nucleus accounts for the mass of an atom.
7. Both protons and neutrons possess the same magnitude of charge but opposite signs.
8. The maximum number of electrons in the L shell is two.
9. An atom is electrically neutral.
10. In phosphide ion, the total number of electrons is equal to 12.

Directions for questions from 11 to 20
Fill in the blanks.

11. The symbol of cobalt is _______.
12. The mass of an electron is _______ times that of mass of a proton.
13. The maximum number of electrons that can be accommodated in the outermost orbit is equal to _______.
14. The valency of argon is _______.
15. In the formation of sodium chloride, each chlorine gains _______ electron.
16. The formula of a zincate radical is _______.
17. The combining capacity of the metal which can exhibit more than one valency is called _______.
18. Blackening of silverware is a _______ change.

19. $2HOCl \rightarrow 2HCl + O_2$ is a _______ decomposition reaction.

20. Atomic number of a metal having 12 neutrons and mass number 23 is _______.

Directions for questions from 21 to 36

For each of the questions, four choices have been provided. Select the correct alternative.

21. The nucleus consists of
(a) proton and electrons.
(b) electrons and neutron.
(c) protons and neutrons.
(d) all the above.

22. The maximum number of electrons that can be accommodated in an orbit is equal to _______ where n is the number of orbits.
(a) $4n^2$ (b) $2n^2$
(c) $(2n + 1)^2$ (d) $n^2 + 4$

23. Among the noble gases which gas does not have an octet configuration?
(a) neon (b) helium
(c) argon (d) xenon

24. Which among the following pairs of elements forms cation and anion, respectively?
(a) sulphur and neon
(b) chlorine and sodium
(c) potassium and phosphorus
(d) aluminium and lithium

25. Which is a negative radical among the following?
(a) silicate (b) ammonium
(c) ferrous (d) chromium

26. Which among the following reactions is not balanced?
(a) $2KClO_3 \rightarrow 2KCl + 3O_2$
(b) $2Fe(OH)3 + 4H_2SO_4 \rightarrow Fe_2(SO_4)_3 + 6H_2O$
(c) $2Pb(NO_3)_2 \rightarrow 2PbO + 4NO_2 + O_2$
(d) $2KNO_3 + FeSO_4 \rightarrow K_2SO_4 + Fe(NO_3)_2$

27. Among the following reactions, identify the synthesis reaction.
(a) $2SO_2 + O_2 \rightarrow 2SO_3$
(b) $C + O_2 \rightarrow CO_2$
(c) $CaO + H_2O \rightarrow Ca(OH)_2$
(d) $2H_2O \rightarrow 2H_2 + O_2$

28. Identify a useful change among the following chemical changes.
(a) blackening of silverware.
(b) rusting of iron.
(c) curdling of milk.
(d) all the above.

29. Which among the following is not a characteristic of a physical change?
(a) The change is temporary.
(b) No new substance is formed.
(c) Molecular formula of a compound changes.
(d) Change of state of matter takes place without any change in molecular composition.

30. Among the following types of chemical reactions, application of metal reactivity series is involved with which reaction?
(a) chemical combination
(b) chemical displacement
(c) chemical decomposition
(d) double decomposition

31. If a solid non-metal X forms oxide of type X_2O_5, then the formula of its corresponding chloride is _______.
(a) XCl_3 (b) XCl_5
(c) X_2Cl_5 (d) X_3Cl_2

32. If the formula of a metal nitrite is $M(NO_2)_2$, then the formula of its dihydrogen phosphate is
(a) $M_2(PO_4)_3$ (b) $MHPO_4$
(c) $M(H_2PO_4)_2$ (d) M_2HPO_4

33. Identify the number of neutrons present in a pair of elements with atomic numbers 20 and 13, if their mass numbers are 40 and 27, respectively.
(a) 20 and 13 (b) 20 and 13
(c) 17 and 20 (d) 20 and 14

34. Heating solid ammonium chloride involves change in the _______.
(a) physical state only.
(b) physical state as well as chemical composition.
(c) colour only.
(d) both colour and volume only.

35. Which of the following is a chemical change?
(a) formation of clouds.
(b) formation of rust.

(c) manufacture of milk powder from milk.
(d) formation of black coating on the walls of a tube light.

36. Arrange the following sequence in order to derive the formula of zinc oxide.
(a) The valencies of ions formed by zinc and oxygen are written below without positive and negative sign.
(b) The positive ion, i.e., Zn^{+2} followed by the negative ion, i.e., oxide (O^{-2}) ion are written with their charges.
(c) The valencies of zinc and oxide ion have common factors, and hence, they are divided by their highest common factor.
(d) The magnitude of valencies of the ions should be interchanged and written as subscripts on right side of the respective ions.
(a) badc (b) abcd
(c) bacd (d) cabd

Directions for questions from 37 to 39

Match the entries given in Column A with the appropriate ones in Column B. More than one suitable content of Column B can also be considered correct for an entry in Column A and some content given in Column B may not be the appropriate one for any entry in Column A.

37.

Column A		Column B
A. $4SO_2$	( )	a. Three atoms of oxygen
B. $2Na^+$	( )	b. Three oxygen molecules
C. $3O_2$	( )	c. Eight molecules of chlorine
D. $8Cl_2$	( )	d. Eight chlorine atoms
	( )	e. Two sodium atoms
	( )	f. Four sulphur dioxide molecules
	( )	g. Two sodium ions

38.

Column A Names of the radical		Column B Charges
A. Aluminium	( )	a. –1
B. Stannous	( )	b. +3
C. Bisulphite	( )	c. +2
D. Silicate	( )	d. +4
	( )	e. –2

39.

Column A		Column B
A. $2Pb(NO_3) \rightarrow PbO + 4NO_2 + O_2$	( )	a. Combination reaction
B. $Zn + FeSO_4 \rightarrow ZnSO_4 + Fe$	( )	b. Double decomposition reaction
C. $NaCl + AgNO_3 \rightarrow NaNO_3 + AgCl$	( )	c. Decomposition reaction
D. $S + O_2 \rightarrow SO_2$	( )	d. Displacement reaction

Level 2

Directions for questions from 1 to 20

For each of the questions, four choices have been provided. Select the correct alternative.

1. X and Y are non-metals in which the M shell is valence shell. Y is precedent to X. X shows uninegative valence. Identify the formulae of the compounds formed when X and Y react with the divalent metal M.
(a) MX_2, MY (b) MX, MY_2
(c) M_3X_2, M_2Y_3 (d) M_2X_3, M_3Y_2

2. Identify the formula of the corresponding hydride of a non-metal X which attains an octet configuration by gaining three electrons.
(a) XH (b) XH_2
(c) XH_3 (d) X_2H_4

3. The number of positively-charged particles in the elements A, B, C are 10, 18, 8, respectively. Identify the element which is chemically reactive?
(a) A (b) B
(c) C (d) both A and B

PRACTICE QUESTIONS

4. If a dispositive ion of an element X consists of 12 protons and 12 neutrons and another dispositive ion of an element Y consists of 18 electrons and twice the number of electrons of dispositive ion X is the number of neutrons of Y, then, the mass number would be in the ratio of
   (a) 1 : 2 (b) 3 : 5
   (c) 12 : 19 (d) 2 : 3

5. If a uninegative ion has 10 neutrons and 9 protons, then the electronic configuration of the atom of the element is
   (a) 2, 8, 1 (b) 2, 7
   (c) 2, 8 (d) 2, 8, 8

6. Mass of total positive charge present in an atom is 55110 times to that of mass of electron. Write the electronic configuration of the element.
   (a) 2, 8, 9, 2 (b) 2, 8, 18, 1
   (c) 2, 8, 18, 2 (d) 2, 8, 8, 2

7. The ratio of difference in the number of electrons between the K and M shells is 3 : 2 for X and Y elements, respectively. X possesses stable electronic configuration and M is the valence shell for both X and Y. Give the electronic configurations of X and Y.
   (a) 2, 8, 5 and 2, 8, 4
   (b) 2, 8, 8 and 2, 8, 6
   (c) 2, 8, 8 and 2, 6, 6
   (d) 2, 8, 6 and 2, 8, 4

8. $2KMnO_4 + H_2SO_4 + H_2C_2O_4 \rightarrow K_2SO_4 + 2MnSO_4 + 10CO_2 + 8H_2O$ Potassium permanganate oxidises oxalic acid in an acidic medium as per the above chemical reaction. Identify the coefficients of H_2SO_4 and $H_2C_2O_4$ in the above equation.
   (a) 3, 5 (b) 4, 4
   (c) 5, 3 (d) 4, 2

9. Heating of a mixture consisting of potassium nitrate, carbon and sulphur involves which of the following chemical changes?
   (a) decomposition and combination
   (b) double decomposition
   (c) combination and displacement
   (d) displacement and decomposition

10. $CuSO_4 + X \rightarrow XSO_4 + Cu$;
   $CuSO_4 + Y \rightarrow YSO_4 + Cu$;
   $XSO_4 + Y \rightarrow$ No reaction;
   $CuSO_4 + Z \rightarrow$ No reaction
   Arrange Cu, X, Y and Z in the ascending order of reactivity.
   (a) X, Y, Z, Cu (b) Cu, Y, X, Z
   (c) Cu, X, Z, Y (d) Z, Cu, Y, X

11. A metal M shows variable valencies 2 and 3. Calculate the ratio of oxygen atoms present in their corresponding sulphates.
   (a) 1 : 3 (b) 3 : 1
   (c) 3 : 4 (d) 3 : 2

12. Identify the formula of the corresponding hydride of a non-metal X which attains an octet configuration by gaining two electrons.
   (a) XH (b) XH_3
   (c) XH_4 (d) XH_2

13. The number of valence electrons in the valence shells, 1, 2, 3 of the X, Y, Z elements are 2, 8, 6, respectively. Identify the element which does not exist in monoatomic state.
   (a) X (b) Z
   (c) Y (d) X and Y

14. The number of neutrons is equal to the number of protons present in the nucleus of two elements and their respective dinegative ions possess 8 electrons in the L and M shells, respectively. Then the ratio of mass numbers of the two elements is
   (a) 1 : 2 (b) 2 : 3
   (c) 3 : 4 (d) 2 : 5

15. Tripositive ion of an element has 13 protons and 14 neutrons. Calculate the ratio of number of electrons present in the K and M shells of the neutral atom.
   (a) 2 : 3 (b) 2 : 1
   (c) 1 : 1 (d) 2 : 5

16. The ratio of number of electrons present in four shells of an element is 1 : 4 : 4 : 1, fourth shell is the valence shell. Then mass of total positive charge present in the atom is __________ times to that of mass of electron
   (a) 36,740 (b) 27,555
   (c) 18,370 (d) 55,110

17. The difference in the number of electrons between K and L and L and M shells is the same for an atom of element X (M shell is the valence shell) and the difference value is the number of valence electrons in the M shell of the other element Y. Give the electronic configurations of their stable ions.

(a) 2, 8, 2 and 2, 8, 6 (b) 2, 8 and 2, 8
(c) 2, 8 and 2, 8, 8 (d) 2, 8, 8 and 2, 8, 8

18. The colour change of $CuSO_4$ solution can be observed when it is made to react with
(a) silver (b) silver chloride solution
(c) zinc (d) gold

19. Identify the ratio of the coefficients of the products CuO, NO_2 and H_2O formed respectively when hydrated copper nitrate is thermally decomposed.

$Cu\ (NO_3)_2.6\ H_2O \rightarrow CuO + NO_2 + H_2O + O_2$

(a) 2 : 3 : 1 (b) 1 : 2 : 3
(c) 1 : 1.5 : 1 (d) 3 : 2 : 1

20. Which of the following chemical change is a photolytic decomposition reaction?
(a) Decomposition of HOCl on standing
(b) Formation of hydrogen chloride from their constituent elements
(c) Decomposition of limestone
(d) Both 1 and 2

21. While explaining the modern concept of an atom, Jack asked the teacher 'why are electrons, assumed to be in continuous motion around the nucleus instead of being stationary?' What could be the explanation given by the teacher?

22. A stable neutral atom X contains two completely-filled orbits. Find out its mass number if it consists of 10 neutrons.

23. Mass of total positive charge present in an atom is 7348 times to that of the mass of a negativily charged particle. Find the atomic number of an element. Find out its valency.

24. While coming back from school in the school bus, Bill asked his friend Gill to predict the atoms, X and Y which form a stable dinegative ion and a uninegative ion, respectively. They have same number of electrons in the L shell which is the shell inner to the valence shell. Predict the answer given by Gill.

25. Predict the valencies of the elements with atomic numbers 7, 15, 16, 18, 19 and justify. Also give the formulae of hydrides formed by the above elements if any.

26. In an atom, the L and M shells have the same number of electrons. The N shell has one electron less than the K shell. Explain the formation of compound by this element and oxygen atom.

27. Explain the formation of the compound formed by two elements with Z = 20 and Z = 17.

28. An atom of an element has the electrons distributed in four shells. If the electrons present in I and IV shells is in 1 : 1 ratio, and the ratio of electrons in II and III shells is also 1 : 1. Predict its valency and justify. Write the formulae of (a) nitride (b) silicate (c) chloride.

29. A negative ion of an atom of element X has 18 electrons and 16 protons. Calculate the number of electrons and protons in its corresponding atom. Also calculate the number of neutrons, if the mass number is 2 units more than double the number of protons.

30. If the formula of metal sulphite of a metal M is MSO_3, give the formula of its
(a) phosphate
(b) hydrogen phosphate
(c) chloride

31. In a chemical laboratory, a student broke the thermometer while performing an experiment. Immediately, the mercury which was spilled over was removed by the lab incharge by sprinkling sulphur on it. The teacher asked the students to predict the type of change/reaction associated with the above process. Explain with appropriate reasons.

32. Asha's mother asked her daughter Asha to boil the milk and after some time, she told Usha, Asha's sister, to add a little curd in the same lukewarm milk. Next day, their father while teaching 'transformation of matter' to his daughters wanted to relate the given activity. So, he asked them, 'what type of change was involved in the activities performed by you yesterday?' Predict the answer given by them and justify.

33. What is the valency of iron in the products formed?
(a) when iron is exposed to humid air for a long time?
(b) when a red hot iron is made to react with steam?

34. Give the necessary chemical equations for the following and balance them.
(a) reaction of a red hot iron with steam to give Fe_3O_4
(b) burning magnesium ribbon in an atmosphere of nitrogen
(c) displacement of copper from the solution of its sulphate by zinc

35. In a science fair, a student took two test tubes labelled A and B filled with copper sulphate solution, then he put magnesium ribbon in test tube A and silver foil in test tube B. Predict the observation and justify.

Level 3

1. An atom of an element X has three completely filled shells. The ratio of electrons in 1 and 4 shells is 1 : 1. Find the atomic number of X. Also write the formulae of its bicarbonate, chromate and phosphide.

2. A metal (M) and a non-metal (X) form a compound with the formula M_4X_3. Predict the electronic arrangements of M and X assuming that M has three occupied shells and X has two occupied shells. Also derive the formulae of (a) sulphate, (b) dihydrogen phosphate and (c) hydroxide of ion of M.

3. Two metals A and B possess the same number of electrons in the L shell and they differ by 1 unit in the number of electrons present in valence shell and number of valence electrons is greater in B than A. The ions formed get the configuration of the nearest inert gases namely argon and neon, respectively. Give the probable electronic configurations of the metals corresponding to A and B and also give the formulae of their corresponding nitrite, nitrate, sulphite and bisulphite.

4. Number of atoms on either side of a chemical equation is balanced by keeping appropriate coefficients and not by changing the subscripts of the elements in the formulae. Justify.

5. An atom of an element has two shells with electrons. The valence shell has 2n + 1 electrons where n is the number of electrons in the inner shell. This element can form a radical X with hydrogen and two radicals Y and Z with oxygen, Z has more number of oxygen than Y. Identify the element and the radicals. Write the reaction taking place between solutions of two salts, AgZ and XCl. Identify the type of reaction.

ASSESSMENT TEST

Test 1

Directions for questions from 1 to 15

Select the correct alternative from the given choices.

1. Which of the following involves only a physical change?
   (a) dissolution of lime in water.
   (b) dissolution of carbon dioxide in water at high pressure.
   (c) dissolution of concentrated acid in water.
   (d) electrolysis of water.

2. Arrange the points in sequence to draw the geometrical representation of an atom which is represented as $_{Z}X^{A}$.
   (a) calculation of the number of protons and neutrons.
   (b) identification of the atomic number and mass number of the given element.
   (c) placement of the electrons in shells around the nucleus according to the electronic configuration.
   (d) writing the electronic configuration of the atom.
   (e) protons and neutrons are shown in nucleus.
   (a) baedc (b) bacde
   (c) abced (d) ecbad

3. (a) $H_2 + Br_2 \rightarrow 2HBr$
   (b) $HBr + NaOH \rightarrow NaBr + H_2O$
   (c) $2NaBr + Cl_2 \rightarrow 2NaCl + Br_2$
   (d) $2NaCl \rightarrow 2Na + Cl_2$
   Arrange the above reactions in order as double displacement reaction, combination reaction, decomposition reaction and displacement reaction.

(a) badc (b) bdac
(c) bacd (d) abdc

4. $2KMnO_4 + H_2SO_4 + 5H_2C_2O_4 \rightarrow K_2SO_4 + 2MnSO_4 + 10CO_2 + 8H_2O$

Potassium permanganate oxidises oxalic acid in an acidic medium as per the above chemical reaction. Identify the coefficient of H_2SO_4 in the above equation.

(a) 3 (b) 4
(c) 5 (d) 2

5. What will be the formula of the sulphate of a trivalent metal that is M^{+3}?

(a) $M_2(SO_4)_3$ (b) MSO_4
(c) M_2SO_4 (d) MSO_3

6. Match the element in column A with the electronic configuration given in column B.

Column A		Column B
(A) Chlorine	( )	(a) 2, 2
(B) Scandium	( )	(b) 2, 8, 7
(C) Manganese	( )	(c) 2, 8, 9, 2
(D) Beryllium	( )	(d) 2, 8, 13, 2

(a) A → c, B → d, C → a, D → b
(b) A → d, B→ a, C → b, D → c
(c) A → b, B → c, C → d, D → a
(d) A → d, B → a, C → c, D → b

7. **Assertion (A)**: Mass number of carbon is 12.

Reason (R): Carbon possesses equal number of all three fundamental particles.

(a) Both A and R are true and R is the correct explanation for A.
(b) Both A and R are true but R is not the correct explanation for A.
(c) A is true and R is false.
(d) A is false and R is true.

8. Identify the atomic number of the element which possesses same number of electrons in the L and M shells.

(a) 18 (b) 19
(c) 16 (d) 18

9. If X^- ion has 10 neutrons and 9 protons, then the electronic configuration of the atom of the element X is

(a) 2, 8, 1 (b) 2, 7
(c) 2, 8 (d) 2, 8, 8

10. **Assertion (A)**: Copper exhibits variable valency.

Reason (R): Copper loses two electrons from its valence shell and it can also lose one more electron from the second last shell.

(a) Both A and R are true and R is the correct explanation for A.
(b) Both A and R are true but R is not the correct explanation for A.
(c) A is true and R is false.
(d) A is false and R is true.

11. The ratio of difference in the number of electrons between the K and M shells is 3 : 2 for X and Y elements, respectively. X possesses stable electronic configuration and M is the valence shell for both X and Y.

The electronic configurations of X and Y are, respectively.

(a) 2, 8, 5 and 2, 8, 4 (b) 2, 8, 8 and 2, 8, 6
(c) 2, 8, 8 and 2, 6, 6 (d) 2, 8, 6 and 2, 8, 4

12. Which of the following is not a chemical change?

(a) passing of steam over a red hot coke.
(b) change in the colour of dil nitric acid upon long standing.
(c) absorption of moisture by P_2O_5.
(d) absorption of moisture by $CaCl_2$.

13. The atomic numbers of three elements A, B and C are 10, 18 and 8, respectively. Identify the element which is chemically reactive?

(a) A (b) B
(c) C (d) both A and B

14. If a dispositive ion of an element X consists of 12 protons and 12 neutrons and another dispositive ion of an element Y consists of 18 electrons and twice the number of electrons of dispositive ion X is the number of neutrons of Y. Then the mass numbers would be in the ratio of

(a) 1 : 2 (b) 3 : 5
(c) 12 : 19 (d) 2 : 3

15. The formula of chloride of a metal is MCl_2. What is the valency of the metal?

(a) 4 (b) 3
(c) 1 (d) 2

Test 2

Directions for questions from 1 to 15

Select the correct alternative from the given choices.

1. Heating solid ammonium chloride involves change in the ________.
   (a) physical state only.
   (b) physical state as well as chemical composition.
   (c) colour only.
   (d) both colour and volume only.

2. Arrange the following points in a sequence to draw the geometrical representation of ${}_{20}Ca^{40}$.
   (a) Protons and neutrons are shown in the nucleus.
   (b) Writing the electronic configuration of the atom.
   (c) Arrange the electrons in shells around the nucleus following the electronic configuration.
   (d) Calculation of the number of protons and neutrons.
   (e) Identification of the atomic number and mass number of the atom.

   (a) edcba (b) edabc
   (c) abcde (d) cabde

3. Four types of reactions are given below in a particular order.
   (a) displacement reaction
   (b) combination reaction
   (c) double displacement
   (d) decomposition reaction

   Follow this sequence to arrange the following reactions:
   (a) $A + B \rightarrow AB$.
   (b) $AB \rightarrow A + B$
   (c) $AB + C \rightarrow CB + A$
   (d) $AC + BD \rightarrow AD + BC$

   (a) cadb (b) bacd
   (c) bdac (d) acbd

4. Identify the coefficients of the products CuO and NO_2 formed respectively when hydrated copper nitrate is thermally decomposed.

   $2Cu(NO_3)_2 . 6\,H_2O \rightarrow CuO + NO_2 + 6H_2O + O_2$

   (a) 3, 4 (b) 2, 4
   (c) 4, 5 (d) 2, 3

5. If the formula of a metal ion is M^{+2}, then the formula of its dihydrogen phosphate is
   (a) $M_2(PO_4)_3$
   (b) $MHPO_4$
   (c) $M(H_2PO_4)_2$
   (d) M_2HPO_4

6. Match the elements in column A with the valency exhibited by the element given in column B.

Column A		Column B		
(A)	Neon	( )	(a)	4
(B)	Oxygen	( )	(b)	3
(C)	Potassium	( )	(c)	2
(D)	Boron	( )	(d)	1
(E)	Silicon	( )	(e)	0

   (a) A → b B → c, C → d, D → e, E → a
   (b) A → b, B → c, C → a, D → e, E → d
   (c) A → e, B → d, C → a, D → b, E → c
   (d) A → e, B → c, C → d, D → b, E → a

7. **Assertion (A)**: The total mass of the protons of a boron atom is 6185 times the total mass of electrons.

   Reason (R): Mass of one proton is 1837 times the mass of electron.
   (a) Both A and R are true and R is the correct explanation for A.
   (b) Both A and R are true but R is not the correct explanation for A.
   (c) A is true and R is false.
   (d) A is false and R is true.

8. Identify the number of neutrons present in a pair of elements with atomic numbers 20 and 13, if their mass numbers are 40 and 27, respectively.
   (a) 20 and 13 (b) 20 and 13
   (c) 17 and 20 (d) 20 and 14

9. Tripositive ion, i.e., X^{+3} has 13 protons and 14 neutrons. Calculate the number of electrons present in the M shell of the neutral atom.
   (a) 2 (b) 1
   (c) 3 (d) 5

10. **Assertion (A)**: Sodium forms a monovalent ion.

    Reason (R): Sodium loses only one electron from the second last shell.

(a) Both A and R are true and R is the correct explanation for A.

(b) Both A and R are true but R is not the correct explanation for A.

(c) A is true and R is false.

(d) A is false and R is true.

11. The difference in the number of electrons between the K and L and L and M shells is the same for an atom of an element X (M shell is the valence shell). And the difference of value is the number of valence electrons in the M shell of the other element Y. The electronic configurations of stable ions of X and Y are, respectively

(a) 2, 8, 2 and 2, 8, 6

(b) 2, 8 and 2, 8

(c) 2, 8 and 2, 8, 8

(d) 2, 8, 8 and 2, 8, 8

12. Which of the following is a chemical change?

(a) formation of clouds

(b) formation of rust

(c) manufacture of milk powder from milk

(d) formation of black coating on the walls of a tube light

13. The atomic numbers of the elements X and Y Z are 2, 10 and 16, respectively. Identify the element which does not exist in monoatomic state.

(a) X (b) Z

(c) Y (d) X and Y

14. The number of neutrons is equal to the number of protons present in the nucleus of two elements X and Y and their dinegative ions possess 8 electrons in the L and M shells, respectively. Then the ratio of mass numbers of the two elements is

(a) 1 : 2 (b) 2 : 3

(c) 3 : 4 (d) 2 : 5

15. If a solid non-metal 'X' forms an oxide of type X_2O_5, then the valency of X is

(a) 3 (b) 5

(c) 6 (d) 4

TEST YOUR CONCEPTS

Fill in the blanks

1. Ca
2. losing two
3. duplet configuration
4. $2SO_3$
5. Ous
6. phosphate
7. Na_2SO_3, Na_2SO_4
8. physical, molecular composition
9. both physical and chemical changes
10. coefficients

Multiple choice questions

11. c	12. b	13. b	14. c	15. b
16. d	17. a	18. b	19. c	20. b
21. d	22. a	23. b	24. c	25. d
26. b				

Match the following

27. A → e
B → d
C → a
D → g

28. A → e
B → d
C → a
D → b

29. A → b
B → a
C → d
D → c

Level 1

True or false

1. True	2. True	3. False	4. True	5. True
6. True	7. False	8. False	9. True	10. False

Fill in the blanks

11. Co
12. $\frac{1}{1837}$
13. 8
14. zero
15. one
16. ZnO_2^{-2}
17. variable valency
18. chemical change
19. photolytic
20. 11

Multiple choice questions

21. c	22. b	23. b	24. c	25. a
26. b	27. b	28. c	29. c	30. b
31. b	32. c	33. d	34. b	35. b
36. c				

Match the following

37. A → f
B → g
C → b
D → c

38. A → b
B → c
C → a
D → e

39. A → c
B → d
C → b
D → a

Level 2

Multiple choice questions

1. a	2. c	3. c	4. b	5. b
6. c	7. b	8. a	9. a	10. d
11. a	12. d	13. b	14. a	15. a
16. a	17. c	18. c	19. b	20. a

ASSESSMENT TEST

Test 1

1. c	2. a	3. a	4. a	5. a
6. c	7. b	8. b	9. b	10. c
11. b	12. d	13. c	14. b	15. d

Test 2

1. b	2. b	3. a	4. b	5. c
6. d	7. d	8. d	9. c	10. c
11. c	12. b	13. b	14. a	15. b

CONCEPT APPLICATION

Level 1

True or false

1. The electrons enter the third shell after the second shell gets completely filled by eight electrons.

2. Helium possesses stable duplet configuration with 2 electrons in the first shell. Hence, its valency is zero.

3. $Cr_2O_7^{2-}$ is called dichromate ion.

4. Rusting of iron is a chemical change in which iron gets corroded in humid air.

5. ublimation of camphor is a physical change.

6. Nucleus consists of both protons and neutrons and it accounts for the mass of the atom.

7. Neutrons carry no charge, i.e., they are neutral, but all protons are positively charged.

8. The maximum number of electrons in the L shell is eight.

9. An atom is electrically neutral due to the presence of equal number of positively and negatively charged particles.

10. P^{3-} = 15 + 3 = 18 electrons

Fill in the blanks

11. The symbol of cobalt is Co.

12. The mass of electron is $\frac{1}{1837}$ times that of mass of a proton.

13. Eight is the maximum number of electrons that can be accommodated in the valence shell.

14. The valency of argon is zero.

15. Chlorine gains an electron while sodium loses an electron to form NaCl.

16. The formula of zincate radical is ZnO_2^{-2}

17. Certain metals like Fe, Cu, Sn, etc., exhibit more than one valency in their compounds called variable valency.

    Examples: $FeCl_2$, $FeCl_3$

18. Blackening of silverware is a chemical change.

19. $2HOCl \rightarrow 2HCl + O_2$ is photolytic decomposition reaction.

20. Atomic number = mass number − number of neutrons

    Atomic number = 23 − 12 = 11

Multiple choice questions

21. The protons and neutrons are present in nucleus.

22. The maximum number of electrons that can be accommodated in an orbit is equal to $2n2$.

23. Helium has a duplet configuration.

24. Potassium loses one electron to form cation and phosphorous gains three electrons to form anion.

25. Silicate is a negative radical, i.e., SiO_3^{-2}.

26. Second reaction is wrongly balanced. Correct balancing is as follows:

    $$2Fe(OH)_3 + 3H_2SO_4 \rightarrow Fe_2(SO_4)_3 + 6H_2O$$

27. $C + O_2 \rightarrow CO_2$ is an example of synthesis reaction.

28. Curdling of milk results in the formation of curd which is a useful change.

29. In a physical change, molecular composition, i.e., molecular formula remains the same.

30. In displacement reactions, application of metal reactivity series is involved.

31. $+5 - X_2O_5$

    Since the valency of chloride ion is 1, the formula of the above chloride would be

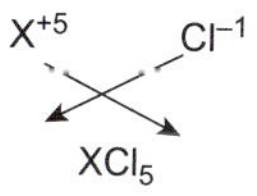

32. $M(NO_2)_2$, hence, valency of metal is 2.

M^{+2} $H_2PO_4^{-1}$

$M(H_2PO_4)_2$

33. $A - Z = n$

(1) $40 - 20 = 20$

(2) $27 - 13 = 14$

34.

$$NH_4Cl(s) \xrightarrow{\Delta} NH_4Cl(g) \xrightarrow{\Delta} NH_3(g) + HCl(g)$$

It involves both sublimation (physical change) and chemical decomposition (chemical change)

35. Formation of rust is a chemical change. When iron is exposed to humid atmosphere, it forms hydrated ferric oxide ($Fe_2O_3 . xH_2O$).

36. (i) The positive ion, i.e., Zn^{+2} followed by the negative ion, i.e., O^{-2} are written with their charges.

(ii) The valencies of ions formed by zinc and oxygen and are written below without positive and negative signs.

(iii) The valencies of zinc and oxide have common factors, and hence, they are divided by the highest common factor.

(iv) The magnitude of the valencies of the ions should be interchanged and written as sub-scripts on right side of the respective ions.

Match the following

37. A → f $4SO^2$ represents four sulphur dioxide molecules

B → g $2Na^+$ represents two sodium ions

C → b $3O_2$ represents three molecules of oxygen

D → c $8Cl_2$ represents 8 molecules of chlorine

38. A → b Charge on aluminium ion is +3

B → c Charge on of stannous ion is +2

C → a Charge on of bisulphate is –1

D → e Charge on silicate ion is –2

39. A → c Lead nitrate is decomposed into PbO, NO_2 and O_2

B → d Zinc displaces iron from $FeSO_4$

C → b Mutual exchange of ions takes place when NaCl and $AgNO_3$ react

D → a Sulphur combines with oxygen to form sulphur dioxide

Level 2

1. Electronic configuration of X is 2, 8, 7

Electronic configuration of Y is 2, 8, 6

∴ Valencies of X and Y are 1 and 2, respectively.

$M^{+2}X^-$ $M^{+2}Y^{-2}$

MX_2 MY

2. X should possess five valence electrons since its valency is 3.

Hence, its corresponding hydride is $X^{-3}\ H^{+1} \Rightarrow XH_3$

3.

Element	'Z'	Electronic configuration
A	10	2, 8
B	18	2, 8, 8
C	8	2, 6

Hence, C is chemically reactive. A and B are chemically inert because of stable octet configuration.

4. X^{+2} = No. of protons = 12 (Z)

Neutrons = 12

No. of electrons = 10

$A = 12 + 12 = 24$

Y^{+2} = 18 electrons

Y = 20 electrons or 20 protons (Z)

$A = 20 + 20 = 40$

$A_x : A_y$

24 : 40

3 : 5

5. X^- = 9 protons (Z)

10 neutrons

10 electrons

$$X \xrightarrow{+1e^{-1}} X^{-1}$$

(9 electrons) (10 electrons)

Therefore, the electronic configuration of an atom is 2, 7.

6. We know that mass of a proton is 1837 times that of an electron.

$\therefore$ Atomic number of the element $= \dfrac{55110}{1837} = 30$

$\therefore$ Electronic configuration of the element is 2, 8, 18, and 2

7. X : Y

3 : 2

2, 8, 8   2, 8, x

$\dfrac{8-2}{x+2} = \dfrac{3}{2}$; X = 6

$\therefore X \rightarrow$ 2, 8, 8 and $Y \rightarrow$ 2, 8, 6

8. $2KMnO_4 + 3H_2SO_4 + 5H_2C_2O_4 \rightarrow K_2SO_4 + 2MnSO_4 + 10CO_2 + 8H_2O$

9. Gunpowder is a mixture of KNO_3, sulphur and coke. When it is heated, KNO_3 undergoes decomposition to produce O_2 and KNO_2. S and C burn in the presence of O_2 producing SO_2 and CO_2.

$$2KNO_3 \xrightarrow{\Delta} 2KNO_2 + O_2$$

$$S + O_2 \rightarrow SO_2$$

$$C + O_2 \rightarrow CO_2$$

Hence, the type of chemical changes involved is decomposition and combination.

10. $CuSO_4 + X \rightarrow XSO_4 + Cu$

$\therefore$ X is more reactive than Cu

$CuSO_4 + Y \rightarrow YSO_4 + Cu$

$\therefore$ Y is more reactive than Cu

$XSO_4 + Y \rightarrow$ No reaction

$\therefore$ X is more reactive than Y

$CuSO_4 + Z \rightarrow$ No reaction

$\therefore$ Z is less reactive than Cu

$\therefore$ order is Z < Cu < Y < X

11. Let the metal be X. The compounds possible are XSO_4 and $X_2(SO_4)_3$. The number of oxygen atoms present is 4 and 12, respectively. Ratio of oxygen atoms is 4 : 12 = 1 : 3

12. X attains an octet configuration after gaining two electrons. Hence, the valency is 2 and valency of hydrogen is 1. $X^{-2}H^{+1} \Rightarrow XH_2$

13.

Element	X	Y	Z
Valence shell	K	L	M
	(1)	(2)	(3)
No. of valence electrons	2	8	6
Elements	He	Ne	S

Sulphur does not exist in monoatomic state due to its greater chemical reactivity.

14. The electronic configuration of the of the dinegative ion with 8 electrons in valence shell (L shell) is

X^{-2} → 2, 8

X → 2, 6 (Z), mass number is 16,

Y^{-2} → 2, 8, 8

Y → 2, 8, 6 (Z)

Mass number is 32 (A) for Y.

Ratio of mass numbers = 16 : 32; 1 : 2

15. Electronic configuration of neutral atom is

K	L	M
2	8	3

Therefore, the ratio of the electrons present in the K and M shells is 2 : 3.

16. The electronic configuration of the element is

K	L	M	N
2	8	8	2

Therefore, the atomic number is 20. We know that mass of proton is 1837 times that of electron.

Therefore, total positive charge present in nucleus is 20 × 1837 = 36740.

17. (X) K L M (Y) K L M

(X) K	L	M	(Y) K	L	M
2	8	X	2	8	6

$\dfrac{8-2}{8-X} = \dfrac{1}{1}$

$8 - X = 8 - 2$

$X = 2$

Stable ion of X is X^{+2} and of Y is Y^{-2}.

X^{+2} : 2, 8 Y^{-2} : 2, 8, 8

18. $CuSO_4 + Zn \rightarrow ZnSO_4 + Cu$

blue colourless

19. $2Cu(NO_3)_2.6H_2O \rightarrow 2CuO + 4NO_2$

$$+ O_2 + 6H_2O$$

Coefficients of CuO, NO_2 and H_2O are respectively 2, 4 and 6.

∴Ratio is 1 : 2 : 3

20. Decomposition of limestone takes place at higher temperatures. Hence, it is a thermal decomposition reaction. Formation of HCl takes place in presence of light. But it is not a decomposition reaction. HOCl on standing in presence of light decomposes to give HCl and O_2.

21. Electrons are negatively charged particles. The nucleus being positively charged exerts a force of attraction on the electrons. In order to counter-balance these forces of attraction, the electrons should be in a continuous motion around the nucleus.

22. Electronic configuration of X is 2, 8

A – Z = number of neutrons

A – 10 = 10

∴mass number (A) is = 20

23. Mass of a proton is 1837 times that of electron

Mass of one proton = 1837

How many protons mass is equal to 7348

$$\frac{7348}{1837} = 4$$

Hence, the atomic number is 4. Electronic arrangement is 2, 2. It can lose 2 electrons to attain duplet configuration, and hence, it forms Be^{+2} ion.

Hence, its valency is +2.

24.

	X^{-2}	Y^{-1}
EC	2, 8, 8	2, 8, 8

Electronic configuration of corresponding neutral atoms, X and Y are as follows:

$X \rightarrow 2, 8, 6$ $\quad Y \rightarrow 2, 8, 7$

Atomic numbers of X and Y are 16 and 17 respectively, and hence, they are sulphur and chlorine.

25.

Atomic number	Element	Electronic configuration	Valency
7	Nitrogen	2, 5	3
15	Phosphorus	2, 8, 5	3
16	Sulphur	2, 8, 6	2
18	Argon	2, 8, 8	0
19	Potassium	2, 8, 8, 1	1

Since N and P need three more electrons to attain an octet configuration, their valencies are 3.

Similarly, the valencies of other elements with atomic numbers 16, 18, 19 are 2, 0 and 1, respectively.

Except argon, all other elements form hydrides.

Phosphorus – PH_3, Potassium – KH

Nitrogen – NH_3

Sulphur – H_2S

Argon does not form any compound

26. The electronic arrangement of the element could be 2, 8, 8, 1. It corresponds to Z = 19 and the element is potassium. Oxygen (Z = 8) has the electronic arrangement of 2, 6. Potassium loses one electron to form K^+ ion, oxygen requires 2 electrons to complete its octet. Therefore, it takes 2 electrons from two potassium atoms and forms K_2O.

27.

Elements	Atomic numbers	Electronic configurations
Ca	20	2, 8, 8, 2
Cl	17	2, 8, 7

Ca loses two electrons to attain an octet configuration while Cl gains an electron to attain the same.

Hence, the formula of the compound is

$Ca^{+2}\ Cl^{-1}$

$CaCl_2$

28.

K	L,	M	N
2,	8	8	2

Hence, the valency of an atom of an element (M) with above electronic configuration is 2 and by losing two electrons it attains an octet configuration.

(a) Formula of its nitride is $M^{+2}N^{-3} \rightarrow M_3N_2$

(b) Formula of its silicate is $M^{+2}SiO_3^{-2} \rightarrow MSiO_3$

(c) Formula of its chloride is $M^{+2}Cl^{-} \rightarrow MCl_2$

29. 16 protons mean that the atomic number of the element (X) is 16. This signifies that the electronic configuration is 2, 8, 6. It has to gain two electrons to attain an octet configuration. Hence, the dinegative ion possesses 18 electrons.

$(X^{-2}) = 16 + 2 = 18.$

A – Z = number of neutrons

Mass number = 2 + 2 (number of protons)

= 2 + 2 (16) = 34

∴ Number of neutrons = 34 – 16 = 18

Hence, number of neutrons = 18

30. In MSO_3, the valency of sulphite is 2. Hence, the valency of M is also 2.

(a) M^{+2} PO_4^{-3}

$M_3(PO_4)_2$

(b) M^{+2} HPO_4^{-2}

$MHPO_4$

(c) M^{+2} Cl^{-1} 1

MCl_2

31. When mercury gets spilled on the floor or any surface, it can be collected by sprinkling sulphur powder because Hg reacts with S to form HgS.

$Hg + S \rightarrow HgS$

It is an element–element combination, i.e., a synthesis reaction.

32. During boiling of milk, the fat particles being lighter than water rise up when some amount of water vaporises. This layer of fat becomes thick and forms a coating over milk upon cooling. Since it involves partial separation of the components of milk, boiling of milk is a physical change. Curdling of milk involves partial conversion of lactose sugar into lactic acid. Since there is change in molecular composition, curdling of milk is a chemical change.

33. (a) When Fe is exposed to humid air for a long period of time, it gets rusted.

Rust: $Fe_2O_3 \times H_2O$

Valency of Fe is 3

(b) $3Fe + 4H_2O \rightarrow Fe_3O_4 + 4H_2$

Steam (ferroso-ferric oxide)

(Fe_3O_4 is a mixed oxide, i.e., $FeO + Fe_2O_3$)

(valency of iron in Fe_3O_4 is 2 as well as 3)

34. (a) $3Fe + 4H_2O \rightarrow Fe_3O_4 + 4H_2$

Steam (ferroso-ferric oxide)

(b) $3Mg + N_2 \rightarrow Mg_3N_2$

(c) $CuSO_4 + Zn \rightarrow ZnSO_4 + Cu$

35.

A	B
$CuSO_4$	$CuSO_4$
+	+
Mg	Ag
↓	↓
$MgSO_4 + Cu$	No reaction

The blue coloured $CuSO_4$ solution in test tube A disappears, and thus, a colourless solution of $MgSO_4$ is formed. This is due to higher reactivity of Mg than Cu. Hence, Mg can displace Cu from $CuSO_4$ solution, and hence, blue-coloured $CuSO_4$ slowly turns colourless. Ag, being less reactive than Cu, cannot displace Cu from $CuSO_4$, and hence, the solution remains blue in colour.

Level 3

1. The electronic arrangement of X must be 2, 8, 18, 2. So, the atomic number of X should be 30. Theelement X is Zn which can form Zn^{+2} ion by losing 2 electrons from the N shell.

Zn^{+2} HCO_3^- (2, 1) → $Zn(HCO_3)_2$

Zn^{+2} CrO_4^{-2} (2, 2 / 1, 1) → $ZnCrO_4$

Zn^{+2} P^{-3} (2, 3) → Zn_3P_2

2. Formula M_4X_3 corresponds to M^{+3} X^{-4}. M has three occupied shells K, L and M. The number of electrons in valence shell is equal to 3. The electronic arrangement could be 2, 8, 3 which corresponds to Z = 13, i.e., aluminium. X has two occupied shells K and L and the L shell must have 4 electrons. The electronic arrangement could be 2, 4 which corresponds to Z = 6, i.e., carbon.

(a) $Al^{+3}SO_4^{-2}$

(b) Al^{+3} $H2PO_4^-$

(c) Al^{+3} OH^-

$Al_2(SO_4)_3$ $Al(H_2PO_4)_3$ $Al(OH)_3$

3.

Shell	A	B
L	8	8
Electronic configuration	2, 8, 8, 1	2, 8, 2
Metal	K	Mg
Formulae	(i) KNO_2	$Mg(NO_2)_2$
	(ii) KNO_3	$Mg(NO_3)_2$
	(iii) K_2SO_3	$MgSO_3$
	(iv) $KHSO_3$	$Mg(HSO_3)$

Or

	A	B
Electronic configuration	2, 8, 8, 2	2, 8, 3
Metal	Ca	Al
Formulae	(i) $Ca(NO_2)_2$	$Al(NO_2)_3$
	(ii) $Ca(NO_3)_2$	$Al(NO_3)_3$
	(iii) $CaSO_3$	$Al_2(SO_3)_3$
	(iv) $Ca(HSO_3)_2$	$Al(HSO_3)_3$

4. If subscripts are changed, then the molecular formulae, i.e., composition of the compound is changed.

Hence, the number of atoms on either side of a chemical equation is balanced by keeping appropriate coefficients.

5. K shell can have only 2 electrons. That means n = 2 and the number of electrons in the L shell is $2n + 1$ which is equal to 5. The electronic arrangement, i.e., 2, 5 corresponds to nitrogen. Nitrogen can form NH_4^+ radical (x) with hydrogen. Y is NO_2^- and Z is NO_3^-.

$AgNO_3 + NH_4Cl \rightarrow AgCl + NH_4NO_4$

This is a double decomposition reaction where silver chloride being insoluble settles at the bottom.

Therefore, this is a precipitation reaction.

ASSESMENT TEST

Test 1

1. Dissolution of any acid in water makes an acid dilute. Hence, it involves only a physical change.

2. (i) Identification of the atomic number and the mass number of the given element.

   (ii) Calculation of the number of protons and neutrons.

   (iii) Protons and neutrons are shown in nucleus.

   (iv) Writing electronic configuration of the atom.

   (v) Placement of the electrons in shells around the nucleus according to the electronic configuration.

3. (i) $HBr + NaOH \rightarrow NaBr + H_2O$

   (ii) $H_2 + Br_2 \rightarrow 2HBr$

   (iii) $2NaCl \rightarrow 2Na + Cl_2$

   (iv) $2NaBr + Cl_2 \rightarrow 2NaCl + Br_2$

4. $2KMnO_4 + 3H_2SO_4 + 5H_2C_2O_4 \rightarrow K_2SO_4 + 2MnSO_4 + 10CO_2 + 8H_2O$

5.

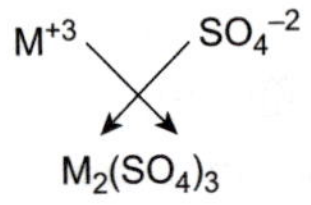

6. A → b Atomic number of chlorine is 17, electronic configuration is 2, 8, 7.

   B → c Atomic number of scandium is 21, electronic configuration is 2, 8, 9, 2.

   C → d Atomic number of manganese is 25, electronic configuration is 2, 8, 13, 2.

   D → a Atomic number of beryllium is 4, electronic configuration is 2, 2.

7. Number of electrons, protons and neutrons in carbon is equal, i.e., 6. The mass number of carbon is 12.

8. Atomic number (Z) — Electronic configuration

Atomic number (Z)	Electronic configuration
19	2, 8, 8, 1
20	2, 8, 8, 2

9. X^- possess 9 protons (Z), 10 neutrons and 10 electrons.

$$\underset{(9\text{ electrons})}{X} \xrightarrow{+1e^{-1}} \underset{(10\text{ electrons})}{X^{-1}}$$

Therefore, the electronic configuration of an atom is 2, 7.

10. Copper loses one electron from its valence shell to form Cu^+ and can loss one more electron from the second last shell to form Cu^{2+} ion.

11. X : Y

2, 8, 8 2, 8, x

$$\frac{8-2}{x-2} = \frac{3}{2} \Rightarrow x = 6$$

$\therefore$ X $\rightarrow$ 2, 8, 8 and Y $\rightarrow$ 2, 8, 6

12. Anhydrous $CaCl_2$, absorbs moisture till it becomes saturated and changes its physical state. Hence, it is a deliquescent substance.

13.

Elements	Atomic numbers	Electronic configurations
A	10	2, 8
B	18	2, 8, 8
C	8	2, 6

Hence, C is chemically reactive. A and B are chemically inert because of stable octet configuration.

14. In X^{+2}, number of protons = 12(Z)

Neutrons = 12

Number of electrons = 10

Mass number, (A) = 12 + 12 = 24

In Y^{+2}, there are 18 electrons.

In Y, there are 20 electrons or 20 protons (Z).

Mass number, (A) = 20 + 20 = 40

$\therefore$ Ratio of mass number = $A_x : A_y$

24 : 40

3 : 5

15. Since the valency of chlorine is 1, the valency of metal MCl_2 is 2.

Test 2

1. $$NH_4Cl_{(s)} \xrightarrow{\Delta} NH_4Cl_{(g)} \xrightarrow{\Delta} NH_{3(g)} + HCl_{(g)}$$

It involves both sublimation (physical change) and chemical decomposition (chemical change)

2. (i) Identification of the atomic number and mass number of the atom.

(ii) Calculation of the number of protons and neutrons.

(iii) Protons and neutrons are shown in the nucleus.

(iv) Writing the electronic configuration of the atom.

(v) Arrange the electrons in shells around the nucleus following the electronic configuration.

3. (i) AB + C $\rightarrow$ CB + A

(ii) A + B $\rightarrow$ AB

(iii) AC + BD $\rightarrow$ AD + BC

(iv) AB $\rightarrow$ A + B

4. $$2Cu(NO_3)_2 \cdot 6H_2O \xrightarrow{\Delta} 2CuO + 4NO_2 + O_2 + 6H_2O$$

Coefficients of CuO and NO_2 are, respectively 2 and 4.

5. Valency of metal is 2.

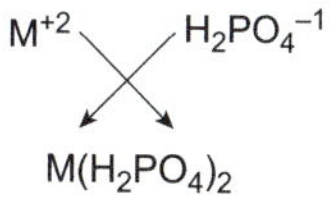

6. A $\rightarrow$ e Atomic number of neon is 10 and electronic configuration 2, 8, valency = 0.

B $\rightarrow$ c Atomic number of oxygen is 8 and electronic configuration 2, 6, valency = 8 – 6 = 2.

C $\rightarrow$ d Atomic number of potassium is 19 and electronic configuration is 2, 8, 8, valency – 1.

D $\rightarrow$ b Atomic number of boron is 5 and electronic configuration is 2, 3, valency = 3.

E → a Atomic number of silicon is 14 and electronic configuration is 2, 8, 4, valency = 4.

7. Mass of one proton is 1837 times the mass of an electron. Hence, the total mass of protons in boron is 9185 times (5 × 1837) the total mass of electrons.

8. A − Z = number of neutrons.

(1) 40 − 20 = 20

(2) 27 − 13 = 14

9. Electronic configuration of neutral atom is

K	L	M
2	8	3

10. Sodium loses only one electron from its outermost shell or valence shell and forms monovalent ion.

11. (X) K L M (Y) K L M

	K	L	M		K	L	M
(X)	2	8	X	(Y)	2	8	6

$$\frac{8-2}{8-X} = \frac{1}{1}$$

8 − X = 8 − 2

$X = 2$

Stable ion of X is X^{+2} and of Y is Y^{-2}.

X^{+2} : 2, 8 Y^{-2} : 2, 8, 8

12. Formation of rust is a chemical change in which iron when exposed to humid atmosphere forms hydrated ferric oxide ($Fe_2O_3 . xH_2O$).

13.

Element	X	Y	Z
Valence shell	K	L	M
	(1)	(2)	(3)
Number of valence electrons			
	2	8	6
Element	He	Ne	S

Sulphur (S) does not exist in monoatomic state due to its greater chemical reactivity.

14. The electronic configuration of the dinegative ion with 8 in electrons in the L valence shell is:

X^{-2} → 2, 8

X → 2, 6 (Z), mass number is 16 for X.

The electronic configuration of the dinegative ion with 8 electrons in the M shell is

Y^{-2} → 2, 8, 8

Y → 2, 8, 6 (Z)

Mass number is 32 (A) for Y.

Ratio of mass number is 16 : 32 ⇒ 1 : 2

15. $\overset{+5}{X_2}\overset{-2}{X_5}$

Valency of non-metal is 5.

HINTS AND EXPLANATION

Chapter 3

Acids, Bases and Salts

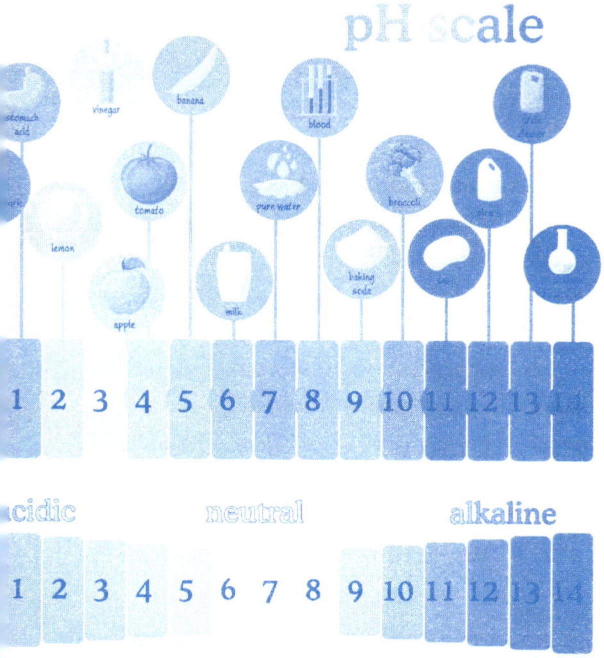

Remember

Before beginning this chapter, you should be able to:

- classify matter on the basis of physical and chemical properties.
- define terms such as atoms, molecules, atomic numbers, etc.
- understand the chemical reactions.

Key Ideas

After completing this chapter, you should be able to:

- define terms such as acids, bases, indicators, acidity, basicity, etc.
- learn about the methods of preparation, properties and uses of acids and bases.
- describe the action of indicators on acids and bases.
- learn about neutralisation, different types of salts, their methods of preparation, properties and uses.

INTRODUCTION

We are very much familiar with the classification of substances as solids, liquids and gases on the basis of their physical state. Similarly, we are aware of the fact that on the basis of chemical composition, the substances are classified as elements, compounds and mixtures. In this chapter, we would be learning about another important categorisation of some chemical compounds into **acids, bases** and **salts**.

It is a familiar experience that some fruits such as oranges, lemons and grapes and other food materials, such as tamarind, have a sour taste. The term, acid, (derived from 'acidus' [Latin] used for expressing sour taste) has been given to the substances which are responsible for the sour taste of the food materials discussed earlier. In the same way, the substances which possess bitter taste like bitter gourd and slippery touch like soaps are called **bases.**

TABLE 3.1 Organic acids

Acids	Sources
Acetic acid	Vinegar
Ascorbic acid	Lemons
Citric acid	All citrus fruits
Tartaric acid	Tamarind, grapes
Lactic acid	Curd
Maleic acid	Grapes
Oleic acid	Olive oil
Stearic acid	Mustard oil
Palmitic acid	Palm oil
Formic acid	Sting of ants and insects
Uric acid	Urine
Oxalic acid	Tomato
Butyric acid	Rancid butter

CLASSIFICATION OF ACIDS BASED ON THE SOURCE OF ORIGIN

Many acids are obtained from the parts of plants and some acids are produced by animals also. These are called **organic acids** due to the fact that their primary sources of origin are living systems. They can also be synthesised in a laboratory (Table 3.1).

The acids which are synthesised from the minerals obtained from the earth's crust are called **mineral acids**.

These mineral acids can be further classified on the basis of various characteristics (Table 3.2).

TABLE 3.2 Mineral acids

S. No.	Chemical names	Common names	Formulae
1.	Hydrochloric acid	Muriatic acid	HCl
2.	Sulphuric acid	Oil of vitriol or king of chemicals	H_2SO_4
3.	Nitric acid	Aqua fortis	HNO_3

Some Common Acids (Mineral Acids or Inorganic Acids) Used in Laboratories

Based on the constituent elements in the acids, they are classified as hydracids and oxyacids. The acids which contain hydrogen and another non-metal are called **hydracids**. The acids which contain oxygen along with hydrogen and non-metal are called **oxyacids.**

Definition of an acid and a base: An acid is defined as a substance which on dissolution in water furnishes H^+ ions as the only positively charged ions. A base is defined as a substance which on dissolution in water furnishes OH^- ions as the only negatively charged ions. The splitting up of a substance into ions when dissolved in water is called ionisation. Various acids and bases differ with respect to the extent of ionisation, and hence, they are classified on the basis of ionisation.

Strong acids and strong bases: The acids and bases which undergo complete ionisation when dissolved in water and furnish large concentration of $\mathbf{H^+}$ or $\mathbf{OH^-}$ ions, respectively are called **strong acids** and **strong bases**.

Examples:

Strong acids

$HCl \rightarrow H^+ + Cl^-$

$H_2SO_4 \rightarrow 2H^+ + SO_4^{-2}$

$HNO_3 \rightarrow H^+ + NO_3^-$

Strong bases

$NaOH \rightarrow Na^+ + OH^-$

$KOH \rightarrow K^+ + OH^-$

Weak acids and **Weak bases:** The acids and bases which ionise partially and furnish less concentration of $\mathbf{H^+}$ and $\mathbf{OH^-}$ ions, respectively are called **weak acids** and **weak bases**. All organic acids are weak acids.

Examples:

Weak acids

$CH_3COOH \rightleftharpoons CH_3COO^- + H^+$

$H_2CO_3 \rightleftharpoons 2H^+ + CO_3^{-2}$

$H_3PO_4 \rightleftharpoons 3H^+ + PO_4^{-3}$

Weak bases

$Ca(OH)_2 \rightleftharpoons Ca^{+2} + 2OH^-$

$Al(OH)_3 \rightleftharpoons Al^{+3} + 3OH^-$

Naming of acids: Hydracids are named as hydro ......ic acid by including the name of the non-metal.

Examples:

HCl	Hydrochloric acid
HBr	Hydrobromic acid
HI	Hydroiodic acid
H_2S	Hydrosulphuric acid

Oxyacids are named based on the number of oxygen atoms present in the acid. Generally, most of the non-metals form two types of oxyacids. The acid containing less number of oxygen atoms is called **ous acid**. The acid containing more number of oxygen atoms is called **ic acid** (Table 3.3).

TABLE 3.3 Naming of acids

Names of non-metal	Formulae of oxy acid	Names of oxy acid
Sulphur	H_2SO_3	Sulphurous acid
	H_2SO_4	Sulphuric acid
Nitrogen	HNO_2	Nitrous acid
	HNO_3	Nitric acid
Chlorine	$HClO_2$	Chlorous acid
	$HClO_3$	Chloric acid
Carbon	H_2CO_3	Carbonic acid
Phosphorus	H_3PO_3	Phosphorous acid
	H_3PO_4	Phosphoric acid

Naming of bases: All oxides of metals are basic in nature. The hydroxides of metals are also basic in nature. Therefore, the bases are named by suffixing oxide term or hydroxide term to the positive radical.

Examples:

$NaOH$ - sodium hydroxide

NH_4OH - ammonium hydroxide

The bases which are soluble in water are called **alkalies**.

Since all the acids and alkalies are soluble in water, the solutions of these substances in water are called **aqueous solutions**. Depending on the relative proportions of water and acid or base present in the solution, the acids and bases are categorised as **concentrated** and **dilute** acids or bases.

Concentrated acids or bases contain lesser proportion of water in contrast to the dilute acids or bases which contain larger proportion of water.

Actual concentration of H^+ or OH^- ions in an aqueous solution of an acid or base respectively depends upon the number of H^+ or OH^- ions furnished by one molecule of an acid or base, respectively on ionisation. This is given by **basicity** of an acid, i.e., the number of H^+ ions furnished by one molecule of acid and **acidity** of a base, i.e., the number of OH^- ions furnished by one molecule of base.

Acids and bases can also be classified on the basis of **basicity** of acids and **acidity** of bases (Tables 3.4 and 3.5).

TABLE 3.4 Classification of acids based on basicity

Types of acid	Basicity	Examples
Monobasic acid	1	$HCl \rightarrow H^+ + Cl^-$
		$HNO_3 \rightarrow H^+ + NO_3^-$
		$CH_3COOH \rightleftharpoons H^+ + CH_3COO^-$
Dibasic acid	2	$H_2SO_4 \rightarrow 2H^+ + SO_4^{-2}$
		$H_2CO_3 \rightleftharpoons 2H^+ + CO_3^{-2}$
		$H_2C_2O_4 \rightleftharpoons 2H^+ + C_2O_4^{-2}$
Tribasic acid	3	$H_3PO_4 \rightleftharpoons 3H^+ + PO_4^{-3}$

TABLE 3.5 Classification of bases based on acidity

Types of acid	Basicity	Examples
Monobasic acid	1	$NaOH \rightarrow Na^+ + OH^-$
		$KOH \rightarrow K^+ + OH^-$
		$NH_4OH \rightleftharpoons NH_4^+ + OH^-$
Dibasic acid	2	$Ca(OH)_2 \rightleftharpoons Ca^{+2} + 2OH^-$
		$Ba(OH)_2 \rightleftharpoons Ba^{+2} + 2OH^-$
Tribasic acid	3	$Al(OH)_3 \rightleftharpoons Al^{+3} + 3OH^-$

Methods of Preparation of Acids and Bases

Acids

1. **By direct combination of elements:** Acids can be prepared by taking two elements like hydrogen gas and chlorine gas. They directly combine and give hydrogen chloride gas (HCl) which dissolves in water and forms hydrochloric acid.

   Example: $H_2 + Cl_2 \rightarrow 2HCl$
   Hydrogen chloride

2. **From non-metals:** Non-metallic elements, on heating with oxygen, give corresponding oxides. Some oxides on dissolution in water give the corresponding acids, and hence, such non-metallic oxides are called **acidic oxides**.

Examples:

$C + O_2 \rightarrow CO_2$

$CO_2 + H_2O \rightarrow H_2CO_3$

$S + O_2 \rightarrow SO_2$

$SO_2 + H_2O \rightarrow H_2SO_3$

$2SO_2 + O_2 \rightarrow 2SO_3$

$SO_3 + H_2O \rightarrow H_2SO_4$

The oxides which are insoluble in water are called **neutral oxides**.

Examples: NO and CO.

Bases

1. **From metals:** Metals on heating with oxygen give metal oxides which are basic in nature. These metal oxides on dissolution in water give corresponding **bases**.

   Examples:

   $4Na + O_2 \rightarrow 2Na_2O$

   $Na_2O + H_2O \rightarrow 2NaOH$

   $2Mg + O_2 \rightarrow 2MgO$

   $MgO + H_2O \rightarrow Mg(OH)_2$

2. **Preparation of slaked lime from quick lime:** When quick lime (calcium oxide) reacts with water, it forms slaked lime (calcium hydroxide) with the liberation of heat.

   $CaO + H_2O \rightarrow Ca(OH)_2 + \text{heat}$
   (Quick lime) (Slaked lime)

Slaked lime is not readily soluble in water and gets settled down at the bottom. The water collected over the slaked lime is called **lime water**.

Physical Properties of Acids and Bases

The physical properties of acids and bases are listed in Table 3.6.

TABLE 3.6 Physical properties of acids and bases

Properties	Acids	Bases
Colour	Mineral acids are colourless liquids but sometimes sulphuric acid becomes yellow due to impurities. Some organic acids are white-coloured solids. **Example**: Benzoic acid	Bases are colourless except the hydroxides of iron and copper
Taste	Sour	Bitter
Touch	–	Slippery
Solubility	Soluble in water	Some bases are soluble in water

Some Important Chemical Properties of Acids

Some of the chemical properties of acids are listed in Table 3.7.

TABLE 3.7 Chemical properties of acids

Reactions	Examples
Reaction with metals: Some metals displace hydrogen from strong acids (HCl, H_2SO_4, etc.). These are called **active metals**	$Zn + H_2SO_4 \rightarrow ZnSO_4 + H_2$ $Mg + 2HCl \rightarrow MgCl_2 + H_2$
Reaction with metal oxides: All dilute mineral acids react with metallic oxides to form their respective metallic salts and water	$ZnO + H_2SO_4 \rightarrow ZnSO_4 + H_2O$ $CaO + 2HNO_3 \rightarrow Ca(NO_3)_2 + H_2O$
Reaction with metal carbonates: This reaction also comes under neutralisation but accompanied by the release of CO_2 gas along with the formation of salt and water	$CaCO_3 + 2HCl \rightarrow CaCl_2 + H_2O + CO_2$ $MgCO_3 + H_2SO_4 \rightarrow MgSO_4 + CO_2 + H_2O$

Some Important Chemical Properties of Bases

Some of the important chemical properties of bases are listed in Table 3.8.

TABLE 3.8 Chemical properties of bases

Reactions	Examples
1. **Action of heat**	NaOH, KOH
Some hydroxides melt on strong heating. They do not undergo decomposition	
2. Other metal hydroxides undergo decomposition on strong heating	$Ca(OH)_2 \xrightarrow{\Delta} CaO + H_2O$ $2Al(OH)_3 \xrightarrow{\Delta} Al_2O_3 + 3H_2O$
3. Ammonium hydroxide decomposes rapidly to give ammonia gas and water upon slight heating	$NH_4OH \xrightarrow{\Delta} NH_3 + H_2O$

Comparative Study of Properties of Acids and Bases

A comparative study of properties of acids and bases is provided in Table 3.9.

TABLE 3.9 Comparative study of properties of acids and bases

Corrosive action on skin: All acids and some alkalies show corrosive action on skin as they form painful blisters when they come in contact with skin.	H_2SO_4 absorbs water from skin tissues. HNO_3 reacts with skin proteins to form a pulp-like mass. NaOH and KOH are called caustic soda and caustic potash, respectively due to their causticizing action on skin.
Reactions taking place between acids and bases: All acids react with alkalies (metal hydroxides) to form salt and water. The reaction of an acid with a base to form salt and water as the products is called **neutralisation**.	$2KOH + H_2SO_4 \rightarrow K_2SO_4 + 2H_2O$ $Ca(OH)_2 + 2HNO_3 \rightarrow Ca(NO_3)_2 + 2H_2O$

EXAMPLE

Why is KOH called caustic potash?

SOLUTION

KOH has causticising action on skin, and therefore, it is called caustic potash.

EXAMPLE

Give an example of dibasic acid. How is it obtained from the corresponding non-metal? Give balanced equations.

SOLUTION

Sulphuric acid is a dibasic acid. Sulphur on heating with oxygen gives sulphur dioxide which on further reaction with oxygen gives sulphur trioxide. SO_3 on dissolution in water forms sulphuric acid.

$S + O_2 \rightarrow SO_2$

$2SO_2 + O_2 \rightarrow 2SO_3$

$SO_3 + H_2O \rightarrow H_2SO_4$

Neutralisation: The reaction between an acid and a base invariably gives salt and water and is called **neutralisation**. In a neutralisation reaction, one H^+ ion of acid is neutralised by one OH^- ion of base. When all the H^+ ions in the acidic solution are neutralised by same number of OH^- ions of basic solution, it is called **complete neutralisation**. The relative amounts of acid and base required for complete neutralisation depend upon the total number of H^+ and OH^- ions produced by the respective acid and base.

Indicators: During the process of neutralisation, a point called **neutralisation point** is reached when all the H^+ ions of an acidic solution are neutralised exactly by the OH^- ions of a basic solution. Further addition of a base to an acid or an acid to a base adds excess OH^- or H^+ ions which changes the nature of the solution from an acidic to a basic or the basic to the acidic, respectively. In order to identify the neutralisation point, during the experiment, certain substances are added which show different colours in acidic and basic solutions. Such substances undergo colour change at the neutralisation point which indicates complete neutralisation and hence are called **indicators**. Certain naturally occurring substances can act as indicators as they change colour based on the nature of a substance.

Example: red cabbage juice.

The other indicators used are litmus, methyl orange and phenolphthalein.

Action of Indicators on Acids and Bases

The action of indicators on acids and bases is shown in Table 3.10.

TABLE 3.10 Action of indicators on acids and bases

Names of indicator	Colours in acidic solution	Colours in basic solution
Litmus	Red	Blue
Methyl orange	Pink (orange red)	Yellow
Phenolphthalein	Colourless	Pink
Turmeric	Yellow	Red

Salt: Salt is a substance which contains positive and negative ions or radicals and ionises completely in its aqueous solution. These are formed by a neutralisation reaction.

Example:

$$\text{Acid} + \text{Base} \xrightarrow{\text{neutralization}} \text{Salt} + \text{Water}$$

When a reaction between acids and bases happens in such a way that some of the replaceable hydrogen ions in acid or replaceable hydroxyl ions in the base are left unreplaced, then these reactions are called **partial neutralisation reactions**. During this process, acidic or basic salts are formed.

Types of Salts

1. **Normal salts:** When all the H^+ ions furnished by an acid molecule are replaced by the metal ions, then the salts formed are called **normal salts**. These salts are formed during complete neutralisation.

   Examples:

   $2NaOH + H_2SO_4 \rightarrow Na_2SO_4 + 2H_2O$

   $2H_3PO_4 + 3Ca(OH)_2 \rightarrow Ca_3(PO_4)_2 + 6H_2O$

2. **Acidic salts**: When a dibasic or tribasic acid is treated with a base and less number of H^+ ions are replaced by the metal ions of the base, then the salt formed is still left with at least one hydrogen molecule. The salt can further undergo neutralisation and eventually forms a normal salt. Such a salt is called an **acidic salt**. These salts furnish hydrogen ions when dissolved in water. These salts are formed by partial neutralisation.

   Examples: $NaOH + H_3PO_4 \rightarrow NaH_2PO_4 + H_2O$
   Sodium dihydrogen phosphate (acidic salt)

   $NaOH + NaH_2PO_4 \rightarrow Na_2HPO_4 + H_2O$
   Disodium hydrogen phosphate (acidic salt)

   $NaOH + Na_2HPO_4 \rightarrow Na_3PO_4 + H_2O$
   Sodium phosphate (normal salt)

3. **Basic salts:** When a diacidic or triacidic base is treated with an acid and less number of OH^- ions are replaced by the negative ion of the acid, then the salt formed is still left with at least one OH^- ion. The salt can further undergo neutralisation to form a normal salt. Such salts are called **basic salts**. These salts furnish hydroxyl ions when dissolved in water. These salts are formed by partial neutralisation.

   Examples: $Ca(OH)_2 + HCl \rightarrow Ca(OH)(Cl) + H_2O$
   Calcium hydroxychloride (basic salt)

   $Ca(OH)Cl + HCl \rightarrow CaCl_2 + H_2O$
   Calcium chloride (normal salt)

Differences Between Acidic Salts and Basic Salts

The differences between an acidic and a basic salt are listed in Table 3.11.

TABLE 3.11 Comparitive study of acidic and basic salts

	Acidic salt	Basic salt
1.	It contains replaceable hydrogen ions.	It contains replaceable hydroxyl ions.
2.	It ionises in water and gives hydrogen ions.	It ionises in water and gives hydroxyl ions.
3.	It turns blue litmus to red.	It turns red litmus to blue.
4.	It gives no colour with phenolphthalein.	It gives pink colour with phenolphthalein.

4. **Double salts**: The salts which contain more than one simple salt are called **double salts**. These salts undergo complete dissociation in their aqueous solutions. They give reactions of all ions present in both the simple salts.

   Examples: (i) Mohr's salt, $FeSO_4 \cdot (NH_4)_2\,SO_4 \cdot 6H_2O$.

   This dissociates into Fe^{2+}, NH_4^+, SO_4^{-2} ions and gives their individual chemical test.

   (ii) Potash alum–, $K_2SO_4 \cdot Al_2(SO_4)_3 \cdot 24H_2O$.

5. **Mixed salts**: The salts which contain more than one basic or acidic radical are called mixed salts. These salts undergo complete dissociation in aqueous solutions. They give the reactions of all the ions produced in aqueous solutions.

Examples: (i) Sodium potassium carbonate ($NaKCO_3$) (Na^+, K^+ and CO_3^{-2} radicals)

(ii) Bleaching powder ($CaOCl_2$) (Ca^{++}, OCl^-, Cl^- radicals)

6. **Complex salts:** The salt which contains one or more simple ions and one complex ion is called **complex salt.**

A complex ion is made up of group of atoms which does not undergo further dissociation into individual atoms or ions and acts as a complete unit.

These salts undergo dissociation into a complex ion and a simple ion.

Examples: (i) Potassium ferrocyanide $K_4[Fe(CN)_6]$

$$K_4[Fe(CN)_6] \xrightarrow{\text{dissociation}} \underset{\text{(Simple ion)}}{4K^+} + \underset{\text{(Complex ion)}}{[Fe(CN)_6]^{4-}}$$

(ii) Sodium argentocyanide $Na[Ag(CN)_2]$

Derivation of names of salts: The name of a salt is basically derived from the acid from which it is formed. The salts derived from the hydracids are given names ending with '**ide**' suffix. If the salt contains hydrogen, then hydrogen is added to the suffix (Table 3.12).

TABLE 3.12 Naming of salts of hydracids

Formula of acids	Formula of bases	Salts formed	Name of the salts
HCl	NaOH	NaCl	Sodium chloride
H_2S	NaOH	Na_2S	Sodium sulphide
H_2S	NaOH	NaHS	Sodium hydrogen sulphide

The salts derived from oxy acids are given names depending on the type of the oxy acid. The salts of '**ous**' acids are given '**ite**' suffix and the salts of '**ic**' acids are given '**ate**' suffix. If hydrogen is present in the salt, hydrogen or bi is added to the name of the radical, i.e., corresponding suffix (Table 3.13).

TABLE 3.13 Naming of salts of oxyacids

Formula of acids	Formula of bases	Formula of salts	Name of the salts
H_2SO_3	KOH	K_2SO_3	Potassium sulphite
		$KHSO_3$	Potassium hydrogen sulphite
H_2SO_4	NaOH	Na_2SO_4	Sodium sulphate
		$NaHSO_4$	Sodium hydrogen sulphate
H_3PO_4	NaOH	$NaPO_4$	Sodium phosphate
		Na_2HPO_4	Disodium hydrogen phosphate
		NaH_2PO_4	Sodium dihydrogen phosphate

In the same way, the salts of nitrous acid and nitric acid are called **nitrites** and **nitrates**, respectively. The salts of carbonic acid are called **carbonates** and **bicarbonates** or **hydrogen carbonates**. The positive ion part is called **basic radical** and the negative ion part is called **acidic radical**.

Preparation of Simple Salts

1. Sodium chloride is prepared by the neutralisation of sodium hydroxide with hydrochloric acid.

$$\underset{\text{(Sodium hydroxide)}}{NaOH} + HCl \rightarrow \underset{\text{(Sodium chloride)}}{NaCl} + H_2O$$

2. $$\underset{\text{Potassium hydroxide}}{KOH} + HCl \rightarrow \underset{\text{Potassium chloride}}{KCl} + H_2O$$

3. $Ca(OH)_2 + 2HCl \rightarrow CaCl_2 + 2H_2O$
Calcium hydroxide   Calcium chloride

4. $ZnO + H_2SO_4 \rightarrow ZnSO_4 + H_2O$
Zinc oxide   Zinc sulphate

5. $CuO + H_2SO_4 \rightarrow CuSO_4 + H_2O$
Copper oxide   Copper sulphate

6. $NH_4OH + HCl \rightarrow NH_4Cl + H_2O$
Ammonium hydroxide   Ammonium chloride

Properties of Salts

1. Salts are solids at room temperature.
2. Salts of sodium and potassium are colourless, whereas salts of iron, copper, chromium, etc., are coloured.

   Examples: Ferrous sulphate is green, aqueous copper sulphate is blue and potassium dichromate is orange red coloured.
3. Salts have high melting points.
4. Most of the salts are soluble in water. The only exceptions are carbonates, sulphides and sulphates of some metals.
5. The aqueous solutions of salts are good conductors of electricity. Hence, they are called electrolytes.
6. When a salt is dissolved in water, depending on the amount of salt dissolved in fixed amount of water, three types of solutions can be obtained.
   (a) **Unsaturated solution**: If the solution is in a state in which it can still dissolve more amount of solute, the solution is said to be **unsaturated**.
   (b) **Saturated solution**: If the solution is in a state in which it cannot dissolve any more amount of solute, the solution is said to be **saturated**.
   (c) **Supersaturated solution**: If more amount of salt is made to dissolve in the saturated solution under some conditions, the solution is said to be **supersaturated**.

Hydrated Salts

In case of some salts, some water molecules are found to be attached to the crystals of salt. These are called **hydrated salts**. In these salts, a fixed number of water molecules gets attached to each molecule of the salt. This is known as **water of crystallisation**. A hydrated salt is represented by writing the number of water molecules along with the formula of the salt.

Examples: (i) $CuSO_4$. 5 H_2O

(ii) $FeSO_4$. 7 H_2O

(iii) Na_2SO_4. $10H_2O$

The salts which do not possess water of crystallisation are called **anhydrous salts**.

Examples: $CuSO_4$, $FeSO_4$, etc.

All hydrated salts lose water of crystallisation on heating and become anhydrous salts. In case of some salts, the colour of the salt is due to the presence of water of crystallisation. Those salts become colourless on heating. For example, $CuSO_4.5H_2O$ is blue in colour and on heating loses blue colour as it forms white-coloured anhydrous $CuSO_4$. However, some salts lose water of crystallisation either completely or partially on exposure to atmospheric air. These are called **efflorescent substances**. For example, $Na_2CO_3.10H_2O$ loses nine water molecules on exposure to air and the remaining one water molecule on heating.

Similarly, there are some substances which absorb moisture from the atmosphere on exposure to air. By absorbing moisture, they may dissolve in the water absorbed and change into solution. Such substances are called **deliquescent substances**.

Examples: Calcium chloride, magnesium chloride, mercuric nitrate, zinc chloride, etc.

However, some substances do not undergo any change in physical state on the absorption of moisture from the atmospheric air. These are called **hygroscopic substances**.

Examples: Calcium oxide, sulphuric acid, cobalt chloride, etc.

Uses of Acids, Bases and Salts

1. **Uses of acids:** Various uses of acids are listed in Table 3.14.

TABLE 3.14 Uses of acids

Name of the acids	Uses	Properties exploited
Hydrochloric acid	Cleaning metal surfaces before tinning and galvanising	1. Reaction of acids with metal oxide coating.
	Dyeing	2. HCl reacts with complex organic compounds and forms coloured compounds.
	Preparation of glucose from starch	3. HCl carries out hydrolysis of starch to glucose.
Sulphuric acid	Manufacture of fertilisers	1. Reaction with bases to form soluble salts.
	Car batteries	2. Electrolyte.
	Dyes, drugs, paints, plastics, detergents	3. Complex organic reactions to form varied products.
Nitric acid	Fertilisers	1. Highly soluble nature of nitrate salts.
	Explosives	TNT, TNG.
	Purification of noble metals (Au, Pt, etc.)	2. Insoluble nature of Au and Pt in HNO_3.
Phosphoric acid	Fertilisers	1. Phosphorus is an essential nutrient for plants.
Carbonic acid	Soda water and soft drinks	1. Tangy taste imparted to water. CO_2 is dissolved in water at high pressure.
Acetic acid	Additive in food stuffs	1. Preservative action and sour taste.
Tartaric acid	Baking powder	1. Neutralising action.

2. **Uses of bases:** Various uses of bases are listed in Table 3.15.

TABLE 3.15 Uses of bases

Name of the bases	Uses	Properties exploited	Solubility in water
Sodium hydroxide (caustic soda) and potassium hydroxide.	Manufacture of soap Paper pulp preparation Synthetic fibres Manufacture of medicines Preparation of metallic sodium Used as a laboratory reagent	1. Reaction with oil 2. Reaction with hydrocarbons of wood	Soluble

Calcium hydroxide	Making mortar Removal of hair from animal skin Spread on beetle leaves Preparation of bleaching powder Used as a good disinfectant in white washing To prepare Bordeaux mixture ($CuSO_4$ + lime)	1. Hardening of mortar on exposure to air. 2. Neutralisation	Slightly soluble
Magnesium hydroxide	Antacid	1. Neutralising action on insoluble excess HCl produced in the stomach.	

3. Uses of salts

Various uses of salts are listed in Table 3.16.

TABLE 3.16 Uses of salts

Chemical names and formulae	Common names	Uses
Sodium chloride, NaCl	Common salt	Food additive Preservative for pickles Manufacture of chlorine and NaOH antacid tablets
Sodium bicarbonate, $NaHCO_3$	Baking soda	Leavening agent for cakes. Fire extinguisher Washing clothes
Sodium carbonate, $Na_2CO_3.10\,H_2O$	Washing soda	Manufacture of glass Detergents
Sodium nitrate, $NaNO_3$	Chile salt petre	Manufacture of nitric acid, gunpowder, and in fireworks Fertiliser
Potassium nitrate, KNO	Nitre	Manufacture of gunpowder, nitric acid Fireworks
Ammonium carbonate, $(NH_4)_2CO_3$	Smelling salt	Relief from common cold
Magnesium sulphate, $MgSO_4.7H_2O$	Epsom salt	Laxative
Ferrous sulphate, $FeSO_4.7H_2O$	Green vitriol	Making of ink, curing of leather
Copper sulphate, $CuSO_4.5H_2O$	Blue vitriol	Fungicide in agriculture, dyeing and printing, electroplating
Calcium sulphate hemihydrate, $CaSO_4.\frac{1}{2}H_2O$	Plaster of Paris	Making of statues, setting of broken bones
$K_2SO_4.Al_2(SO_4)_3.24H_2O$	Potash alum	Purification of water
Ammonium chloride, NH_4Cl	Salammoniac	Used as a laboratory reagent
Calcium oxychloride, $CaOCl_2$	Bleaching powder	Bleaching clothes and purification of water
Sodium nitrate, $NaNO_3$	Chile salt petre	As fertiliser, in glass and fireworks
Sodium thiosulphate, $Na_2S_2O_3$	Hypo	As an antichlor for removing excess of chlorine from the bleached articles in textile industries
Zinc sulphate, $ZnSO_4.7H_2O$	White vitriol	In medicine as an eye lotion and in making lithopone, a paint base

EXAMPLE

Why is nitric acid used in the purification of gold?

SOLUTION

Nitric acid is a good oxidising agent. Gold is purified by treating it with HNO_3; all the impurities are oxidised by nitric acid.

EXAMPLE

What is meant by neutralisation? Give balanced equations showing neutralisation reactions of the following.

(a) Slaked lime and sulphuric acid

(b) Sodium bicarbonate and hydrochloric acid.

(c) Magnesium carbonate and sulphuric acid.

SOLUTION

All acids react with alkalies to form salt and water. The reaction of an acid with a base to form salt and water as the products is called neutralisation.

(a) $Ca(OH)_2 + H_2SO_4 \rightarrow CaSO_4 + 2H_2O$

(b) $NaHCO_3 + HCl \rightarrow NaCl + CO_2 + H_2O$

(c) $MgCO_3 + H_2SO_4 \rightarrow MgSO_4 + CO_2 + H_2O$

POINTS TO REMEMBER

- Strength of acids and bases depends on the extent of ionisation.
- Oxides of metals and non-metals, on hydrolysis, give bases and acids, respectively.
- Metals which are more reactive than hydrogen, on treatment with dilute hydrochloric acid or dilute sulphuric acid, liberate hydrogen.
- Neutralisation is the process in which an acid reacts with a base to form salt and water.
- Indicators are used to identify the neutralisation point.
- Calcium hydroxide is used for white washing and preparation of mortar.
- Acidic salts contain replaceable H^+ ions, basic salts contain replaceable OH^- ions and normal salts contain neither replaceable H^+ nor OH^- ions.
- Aqueous solutions of salts are good conductors of electricity.
- Hydrated salts possess water of crystallisation while anhydrous salts do not possess any water of crystallisation.
-

Nature of salts or substances	Properties exploited
Efflorescent	Lose water of crystallisation on exposure to air
Deliquescent	Absorb moisture on exposure to air and undergo change in state
Hygroscopic	Absorb moisture without undergoing any change in state

-

Oxyacid	Corresponding salt suffix
Sulphuric acid	Sulphate
Sulphurous acid	Sulphite
Hydrochloric acid	Chloride
Nitric acid	Nitrate
Nitrous acid	Nitrite
Phosphoric acid	Phosphate

- Certain salts are well known by their common names, rather than by their chemical names.

Examples:

1. Baking soda is the common name of sodium bicarbonate.
2. Nitre is the common name of potassium nitrate.

TEST YOUR CONCEPTS

Very Short Answer Type Questions

Directions for questions from 1 to 10

Fill in the blanks.

1. The acids which are obtained from the minerals are called __________.
2. The oxides of __________ on hydrolysis give bases.
3. Ammonium hydroxide on heating gives __________ and __________.
4. Zinc on reaction with dilute sulphuric acid liberates __________ gas.
5. Mineral acids react with metallic carbonate to form its respective metallic salt, __________ and __________.
6. Methyl orange turns lime water __________.
7. Chemical formula and chemical name of plaster of Paris are __________ and __________ respectively.
8. The salt formed by removal of water of crystallisation from hydrated salt is called __________.
9. Hydrated zinc sulphate is __________ in colour.
10. __________ which is a hydrated salt is used as a fungicide in agriculture.

Directions for questions from 11 to 28

For each of the questions, four choices have been provided. Select the correct alternative.

11. Which among the following acids is present in lemons?
    (a) acetic acid (b) oleic acid
    (c) stearic acid (d) ascorbic acid
12. Identify the weak mineral acid among the following.
    (a) palmitic acid (b) acetic acid
    (c) carbonic acid (d) hydrochloric acid
13. For the formation of a normal salt from one molecule of H_2SO_4, how many molecules of NaOH are required?
    (a) one (b) two
    (c) four (d) three
14. Which among the following pairs of bases do not decompose on heating?
    (a) calcium hydroxide, potassium hydroxide
    (b) sodium hydroxide, potassium hydroxide
    (c) copper hydroxide, calcium hydroxide
    (d) codium hydroxide, copper hydroxide
15. Which among the following is a strong acid?
    (a) acetic acid (b) sulphurous acid
    (c) carbonic acid (d) nitric acid
16. Among the following, which substance turns phenolphthalein to pink?
    (a) soda water (b) lime water
    (c) common salt (d) sugar solution
17. Washing soda is__________
    (a) hydrated sodium carbonate
    (b) anhydrous sodium carbonate
    (c) hydrated magnesium sulphate
    (d) anhydrous magnesium sulphate
18. A salt formed by the partial neutralisation of hydroxyl ions of a base by an acid is called __________ salt.
    (a) normal (b) acidic
    (c) basic (d) none of these
19. Which among the following is a hydrated salt?
    (a) marble (b) baking soda
    (c) green vitriol (d) all the above
20. Salt used in purification of water is
    (a) potash alum (b) epsom salt
    (c) green vitriol (d) blue vitriol
21. Monobasic weak acid among the following is
    (a) carbonic acid (b) oxalic acid
    (c) acetic acid (d) nitric acid
22. __________ on hydrolysis gives a strong acid.
    (a) sulphur dioxide (b) carbon dioxide
    (c) sulphur trioxide (d) nitrogen oxide
23. Which among the following is the application of neutralisation?
    (a) Farmers add slaked lime to soil.

(b) Persons suffering from acidity are given antacid tablets.
(c) Usage of lithium hydroxide in submarines.
(d) All the above.

24. Salt solutions are good conductors of electricity. This is due to the presence of ________ in solution state.
(a) mobile electrons (b) mobile ions
(c) molecules (d) both 1 and 3

25. Which among the following is used to remove moisture from the surroundings?
(a) calcium chloride
(b) magnesium chloride
(c) zinc chloride
(d) calcium oxide

26. X is a base which is soluble in water. Then, the metal present in X may be
(a) aluminium (b) copper
(c) magnesium (d) manganese

27. A non-metallic element is converted into a compound X after a series of reactions. A little amount of X when tested with blue litmus turns to red. X on complete reaction with another compound Y gave the product which did not respond to litmus test. Identify the correct sequence of the reactions.
(1) $2SO_2 + O_2 \rightarrow 2SO_3$
(2) $H_2SO_4 + 2KOH \rightarrow K_2SO_4 + 2H_2O$
(3) $SO_3 + H_2O \rightarrow H_2SO_4$
(4) $S + O_2 \rightarrow SO_2$
(a) 1342 (b) 4132
(c) 2314 (d) 4321

28. Arrange the pairs in sequence in decreasing order of basicity of an acid and acidity of a base followed by the normal salts formed by the given acid–base pairs in the same order, respectively.
(1) acetic acid and ammonium hydroxide
(2) oxalic acid and barium hydroxide
(3) phosphoric acid and aluminium hydroxide
(4) barium oxalate
(5) ammonium acetate
(6) aluminium phosphate
(a) 14365 (b) 321645
(c) 143526 (d) 341526

***Directions for questions from 29 to 31:* Match the entries given in Column A with the appropriate ones in Column B. More than one suitable content of Column B can also be considered correct for an entry in Column A and some content given in Column B may not be the appropriate one for any entry in Column A.**

29.

Column A	Column B
A. Nitric acid	( ) a. Weak acid
B. Phosphoric acid	( ) b. Strong acid
C. Calcium hydroxide	( ) c. Car batteries
D. Sulphuric acid	( ) d. Bleaching powder
	( ) e. Baking powder

30.

Column A	Column B
A. Sodium dihydrogen phosphate	( ) a. Basic salt
B. Calcium hydroxy chloride	( ) b. Gunpowder
C. Nitre	( ) c. Manufacture of glass
D. Ammonium carbonate	( ) d. Acidic salt
	( ) e. Normal salt
	( ) f. Smelling salt

31.

Column A	Column B
A. Efflorescent	( ) a. Cobalt chloride
B. Deliquescent	( ) b. Sodium chloride
C. Hygroscopic	( ) c. Zinc chloride
D. Food additive	( ) d. Mercuric nitrate
	( ) e. Washing soda

Directions for questions from 32 to 51
Answer the following questions.

32. Name the acid present in
(a) curd (b) tamarind

33. Name the acid which is produced by the sting of an ant.

34. What is meant by an alkali? Give two examples.

35. Give common names of sodium hydroxide and potassium hydroxide.

36. Name the substances formed by the dissolution of following oxides in water.
(a) CO_2 (b) MgO

37. What is the action of litmus on an acid and a base?

38. Name a natural indicator.

39. Give the formulae of the following salts:

(a) blue vitriol (b) green vitriol
(c) potash alum

40. Name the salt used for the manufacture of glass. Give its formula.

41. Name the salts used for the following purposes:
(a) Purification of water (b) Fire extinguisher

42. Name the acid present in
(a) lemon (b) vinegar

43. What are neutral oxides? Give examples.

44. Name the acid and base from which the following salts are formed.
(a) sodium chloride (b) potassium sulphate
(c) calcium nitrate

45. What are the suffixes of salts formed from the following acids?
(a) carbonic acid (b) nitrous acid
(c) sulphuric acid (d) nitric acid

46. Write the chemical formula of hypo.

47. Identify the basic and acidic radicals in sodium nitrate.

48. Give two examples of hydrated salts.

49. Define water of crystallisation.

50. What is the formula of plaster of Paris?

Short Answer Type Questions

52. Distinguish between organic acids and mineral acids. Give examples.

53. Give an example of a dibasic acid. How is it obtained from the corresponding non-metal? Give balanced equations.

54. What are active metals? Give examples with equations.

55. Explain the corrosive action of acids and alkalies. Give examples.

56. Distinguish between an acid and base with respect to definitions.

57. Define the following and give an example for each.
(a) efflorescent substances
(b) deliquescent salts
(c) hygroscopic salts

58. What are mineral acids? How can they be further classified on the basis of nature of the constituents?

59. Distinguish between strong acids and weak acids. Give two examples of each.

60. Write a short note on naming of acids.

61. What is meant by neutralisation? Give balanced equations showing neutralisation reactions of the following:
(a) slaked lime and sulphuric acid
(b) sodium bicarbonate and hydrochloric acid
(c) magnesium carbonate and sulphuric acid

62. What is the action of heat on the following?
(a) ammonium hydroxide
(b) aluminium hydroxide
(c) sodium hydroxide

63. What are indicators? Give examples.

64. Distinguish between acidic salts and basic salts.

65. Mention four important properties of salts.

66. Mention two uses of the salts given below:
(a) sodium chloride
(b) sodium bicarbonate
(c) calcium sulphate hemihydrate

67. Mention two important uses of the following acids and also mention the property exploited.
(a) hydrochloric acid (b) nitric acid

Essay Type Questions

68. Define basicity of an acid and acidity of a base. On the basis of these, explain the classification of acids and bases. Give suitable examples.

69. Write a short note on different types of salts.

70. Give the chemical properties of acids and bases by giving balanced chemical equations.

71. Describe the methods of preparation of acids and bases with the help of examples.

72. Give equations for the dissociation of the following compounds in their solutions and write the formulae of the salts formed from them.
(a) H_2SO_4 (b) $Mg(OH)_2$

73. Explain the naming of salts with examples.

74. (1) By using moist blue litmus paper, classify the following compounds given below as acids and bases:
$Ca(OH)_2$, H_2SO_4, CH_3COOH, NaOH, KOH, $Mg(OH)_2$, HNO_3, NH_4OH
(2) Mention the acidic or basic or neutral nature of solutions by using
(a) litmus
(b) methyl orange
(c) phenolphthalein indicators

75. Mention the uses of the following bases and mention the property exploited in those uses.
(a) sodium hydroxide
(b) calcium hydroxide
(c) magnesium hydroxide

For Hints and Explanations, please visit:
www.pearsoned.co.in/IITFoundationSeries

CONCEPT APPLICATION

Level 1

Directions for questions from 1 to 10
State whether the following statements are true or false.

1. Hydrochloric acid is a monobasic acid.
2. Carbonic acid is a strong acid.
3. Sulphur dioxide on hydrolysis gives sulphuric acid.
4. Acids turn methyl orange solution to pink.
5. Ammonium hydroxide is an alkali.
6. Lead sulphide is a normal salt.
7. Formula of washing soda is Na_2CO_3. $7H_2O$.
8. Nitre is the common name of sodium nitrate.
9. Mercuric nitrate is a deliquescent salt.
10. Chile salt petre is used in the manufacture of nitric acid.

Directions for questions from 11 to 20
Fill in the blanks.

11. Hydrochloric acid is a _______ acid. (monobasic/dibasic)
12. Carbonic acid is a _______ acid. (strong/weak)
13. Sulphur dioxide on hydrolysis gives _______.
14. Acid turns methyl orange solution _______.
15. Ammonium hydroxide is an _______. (acid/alkali)
16. Sodium sulphate is a _______ salt.
17. Formula of washing soda is _______.
18. Nitre is the common name of _______.
19. Mercuric nitrate is a _______ salt. (deliquescent/efflorescent)
20. Chile salt petre is used in the manufacture of _______. (nitric acid/nitrous acid)

Directions for questions from 21 to 38
For each of the questions, four choices have been provided. Select the correct alternative.

21. Identify the organic acid among the following given acids:
(a) hydroiodic acid (b) nitrous acid
(c) carbonic acid (d) stearic acid

22. Identify the strong acid and weak base, respectively, among the following:
(a) HCl and KOH
(b) $Ca(OH)_2$ and $Al(OH)_3$
(c) HNO_3 and $Al(OH)_3$
(d) H_2CO_3 and NH_4OH

23. H_3PO_4 is an example of
(a) monobasic acid (b) dibasic acid
(c) tribasic acid (d) strong acid

24. $Ca(OH)_2$ in an example of
(a) monoacidic base (b) diacidic base
(c) triacidic base (d) strong base

25. $CaCl_2$ is an example of
(a) normal salt (b) acidic salt
(c) basic salt (d) complex salt

26. Among the following, identify an extremely corrosive alkali.
    (a) ammonium hydroxide
    (b) calcium hydroxide
    (c) potassium hydroxide
    (d) magnesium hydroxide

27. Generally, hydrolysis of metal oxides gives
    (a) acids (b) bases
    (c) salts (d) hydrogen gas

28. Name of the salt formed by treating calcium oxide solution with nitric acid is ________.
    (a) calcium nitrate
    (b) calcium nitrite
    (c) calcium hydrogen nitrate
    (d) calcium hydrogen nitrite

29. Formula of potassium hydrogen sulphite is
    (a) $KHSO_3$ (b) K_2SO_3
    (c) K_2SO_4 (d) $KHSO_4$

30. Smelling salt is
    (a) sodium chloride
    (b) calcium carbonate
    (c) ammonium chloride
    (d) ammonium carbonate

31. Dibasic weak mineral acid among the following is
    (a) acetic acid
    (b) sulphuric acid
    (c) phosphoric acid
    (d) sulphurous acid

32. If a non-metal forms only two oxides and one oxide on hydrolysis gives an acid, then the non-metal could be
    (a) sulphur (b) carbon
    (c) phosphorus (d) nitrogen

33. Antacid pair among the following is
    (a) $Mg(OH)_2$, NaOH
    (b) $Mg(OH)_2$, $Al(OH)_3$
    (c) KOH, NaOH
    (d) $Al(OH)_3$, NaOH

34. Which among the following is/are poor conductors of electricity?
    (a) concentrated and pure H_2SO_4
    (b) concentrated and pure HNO_3
    (c) liquid hydrogen chloride
    (d) all the above

35. Hygroscopic substance among the following is
    (a) calcium oxide
    (b) sodium oxide
    (c) common salt
    (d) calcium chloride

36. Among the following, identify the pair of water-soluble bases.
    (a) copper oxide and potassium oxide
    (b) copper oxide and calcium oxide
    (c) magnesium oxide and potassium oxide
    (d) aluminium oxide and magnesium oxide

37. A metallic element is converted into a compound 'A' after a series of reactions. Compound A turns red litmus to blue. A on reaction with another compound B which turns blue litmus to red gave a compound C which responded to litmus test like compound B C on further reaction with A gave a compound D which did not respond to litmus test. Identify the correct sequence of steps for the formation of A, B, C and D.
    (1) $NaOH + H_2SO_4 \rightarrow NaHSO_4 + H_2O$
    (2) $4Na + O_2 \rightarrow 2Na_2O$
    (3) $NaHSO_4 + NaOH \rightarrow Na_2SO_4 + H_2O$
    (4) $Na_2O + H_2O \rightarrow 2NaOH$
    (a) 1324 (b) 2314
    (c) 2413 (d) 4213

38. Arrange the following pairs in a sequence of decreasing order of acidity of base and increasing order of basicity of an acid and the salts formed between an acid and base with equal basicity and acidity in their increasing order, respectively.
    (1) sodium chloride
    (2) barium hydroxide and sulphuric acid
    (3) ferric phosphate
    (4) ferric hydroxide and hydrochloric acid
    (5) barium sulphate
    (6) sodium hydroxide and phosphoric acid
    (a) 426153 (b) 624153
    (c) 426351 (d) 624351

Directions for questions from 39 to 41

Match the entries given in Column A with the appropriate ones in Column B. More than one suitable content of Column B can also be considered correct for an entry in Column A and some content given in Column B may not be the appropriate one for any entry in Column A.

39.

Column A	Column B
A. Carbonic acid	( ) a. Baking powder
B. Tartaric acid	( ) b. Antacid
C. Caustic soda	( ) c. Soft drinks
D. Milk of magnesia	( ) d. Additive in food stuffs
	( ) e. Soap industry

40.

Column A	Column B
A. Green vitriol	( ) a. Na_2CO_3 $10H_2O$
B. Blue vitriol	( ) b. $MgSO_4$ $7H_2O$
C. Epsom salt	( ) c. $FeSO_4$ $7H_2O$
D. Washing soda	( ) d. $CuSO_4$ $5H_2O$
	( ) e. $CaSO_4$ ½ H_2O

41.

Column A (indicator)	Column B (colour in acidic solution)
A. Methyl orange	( ) a. Red
B. Phenolphthalein	( ) b. Yellow
C. Turmeric	( ) c. Colourless
D. Litmus	( ) d. Blue
	( ) e. Pink

Level 2

Directions for questions from 1 to 18

For each of the questions, four choices have been provided. Select the correct alternative.

1. Odd one among the following with respect to the strength of acids is
   (a) phosphoric acid (b) carbonic acid
   (c) sulphuric acid (d) acetic acid

2. One molecule of an 'ic' acid of a non-metal having 5 electrons in valence shell reacts with a molecule of base to form a salt X. The base corresponds to the metal with one electron in valence shell. If the salt so formed can react with the same base in 1 : 2 ratio, predict the formula of the salt X.
   (a) K_2SO_4 (b) Na_2HPO_4
   (c) KH_2PO_4 (d) Na_3PO_4

3. Three elements form their respective oxides A, B and C. A and C are gases and B is a solid which on dissolution in water turns red litmus to blue. In presence of moisture, A turns blue litmus to red and C is neutral to litmus. Then, A, B and C may be, respectively
   (a) SO_3, MgO, NO
   (b) SO_2, CaO, CO_2
   (c) CO,Na_2O, SO_3
   (d) SO_2, $Ca(OH)_2$, NO_2

4. Two salts X and Y are taken in two test tubes A and B, respectively, and subjected to heating. Water is added to two test tubes, in case of A salt regains its original colour and in case of B water starts boiling. Then, X and Y may be, respectively
   (a) blue vitriol and limestone
   (b) blue vitriol and baking soda
   (c) nitre and limestone
   (d) nitre and washing soda

5. The salt formed by complete neutralisation of calcium hydroxide with oxy acid of sulphur having four oxygen atoms is
   (a) calcium sulphite (b) calcium bisulphate
   (c) calcium sulphate (d) calcium bisulphite

6. Role of nitre in the manufacture of gunpowder is ________.
   (a) to supply oxygen
   (b) to supply nitrogen
   (c) to decrease the rate of combustion
   (d) absorb temperature produced by combustion

7. Generally, pickles are stored in
   (a) tin vessels or glass vessels
   (b) tin vessels or plastic vessels
   (c) glass vessels or plastic vessels
   (d) aluminium vessels or tin vessels

8. Pure HNO_3 is colourless, however, it gradually becomes yellow on standing due to _____.
   (a) the decomposition of HNO_3 and formation of NO

PRACTICE QUESTIONS

(b) the decomposition of HNO_3 and formation of NO_2

(c) oxidation by atmospheric O_2 and formation of NO

(d) oxidation by atmospheric O_2 and formation of NO_2

9. H_2SO_4 is used in the manufacture of fertilisers. The property exploited in H_2SO_4 is
   (a) formation of soluble salts
   (b) strong acid
   (c) good electrolyte
   (d) good dehydrating agent

10. Weak acid among the following is
   (a) sulphuric acid (b) hydrochloric acid
   (c) nitric acid (d) carbonic acid

11. A salt 'x' has a positive radical in which the metal has electronic configuration 2, 8, 8, 2. The negative radical is formed from an acid used in soft drinks. One molecule of the salt 'x' reacts with one molecule of a base. Predict the formula of the salt 'x.'
   (a) $Mg(HCO_3)_2$ (b) $Ca(HCO_3)_2$
   (c) $CaCO_3$ (d) $MgCO_3$

12. Which among the following is true regarding aqueous solution of sulphur trioxide and sodium oxide?
   (a) both solutions turn blue litmus to red.
   (b) Both solutions turn red litmus to blue.
   (c) Aqueous solutions of SO_3 and Na_2O turn blue litmus to red and red litmus to blue, respectively.
   (d) Aqueous solutions of SO_3 and Na_2O turn red litmus to blue and blue litmus to red, respectively.

13. Blue coloured solid anhydrous cobalt chloride is exposed to atmosphere and forms a pink-coloured solid. From this, we can infer that cobalt chloride is
   (a) hygroscopic (b) deliquescent
   (c) efflorescent (d) none of these

14. Name and formula of the salt formed by complete neutralisation of aluminium hydroxide with ic acid of sulphur having four oxygen atoms is
   (a) aluminium sulphite, $Al_2(SO_3)_3$
   (b) aluminium sulphate, $Al_2(SO_4)_3$
   (c) aluminium sulphate, $Al_2(SO_3)_3$
   (d) aluminium sulphite, $Al_2(SO_4)_3$

15. Gaseous products formed when gunpowder is subjected to heating are dissolved in water. Identify the products formed on dissolution.
   (a) H_2CO_3, HNO_3 (b) H_2SO_3, H_2SO_4
   (c) H_2SO_4, HNO_2 (d) H_2SO_3, H_2CO_3

16. Generally, pickles are not stored in tin vessels because
   (a) tin is a costly metal.
   (b) tin is having high reactivity towards food components.
   (c) tin is a denser metal.
   (d) tin shows corrosive action.

17. Pure nitric acid on longstanding turns yellow due to the formation of
   (a) O_2, NO and H_2O (b) O_2, NO_2 and H_2O
   (c) NO, NO_2 and H_2O (d) N_2O, O_2 and H_2O

18. Identify the acid used in the purification of metals like gold and silver
   (a) sulphuric acid (b) phosphoric acid
   (c) hydrochloric acid (d) nitric acid

19. Though phosphorus acid contains three hydrogen molecules, it is a dibasic acid. Give reasons.

20. While showing some common compounds used in our daily life in the laboratory, the teacher showed sodium chloride compound and told that it is also called common salt. Next day, one of the students brought common salt from home to the laboratory and requested the teacher to show sodium chloride as well. He kept both the common salt and sodium chloride in two different watch glasses and left. The next day, during lunch break the student went to the laboratory and saw that the common salt brought from home became sticky, whereas the compound sodium chloride remained as it is. What would be the justification given by the teacher for the above observation?

21. Two friends Neela and Leela went to a shop, brought mango and lemon pickle of 1 kg each, respectively. Neela stored the pickle in a tin-coated iron vessel, whereas Leela stored her pickle in a glass vessel. After a few days, Neela found that her pickle got spoiled, whereas Leela's pickle did not. Explain the reason behind it.

PRACTICE QUESTIONS

22. Hydrochloric acid is used to clean the metal surface before the metal surface is subjected to galvanisation. Explain.

23. Why is nitric acid used in the purification of gold?

24. Hydrated sodium sulphate and hydrated copper sulphate are taken in two air-tight containers X and Y, respectively. When blue-coloured $CoCl_2$ is introduced into the two containers, there is colour change in one container. Identify the container and give reasons in support of your answer.

25. A teacher demonstrated action of acids on metals in a laboratory. After the activity was over, one naughty student took an aluminium rod. He dipped it in a bottle containing concentrated nitric acid. Later, he dipped it in a beaker containing dil H_2SO_4. But he did not find any bubbles coming out. He was surprised because he was able to see bubbles coming out when a metal rod is dipped in sulphuric acid when the teacher conducted the experiment. He then approached the teacher and asked what is wrong with his activity. Predict the answer given by the teacher.

26. In a chemistry practical examination, three friends DEAN, BEAN and SEAN were given test tubes A, B, and C, respectively filled with solutions. DEAN added blue litmus, BEAN added methyl orange and SEAN added the litmus solution to the given test tube, respectively. DEAN found the blue litmus turning to red, BEAN observed methyl orange turning to yellow but SEAN found no action in the solution provided to him in the test tube. Comment on the nature of the three solutions given.

27. One molecule of H_3PO_4 is treated with one molecule of NaOH, two molecules of NaOH and three molecules of NaOH separately. Comment on the nature of products formed and write their names.

28. An element X on treatment with oxygen gives Y which on hydrolysis produces Z. Z on treatment with higher oxide of carbon A gives a milky white precipitate B. Identify X, Y, Z, A and B?

29. Element X has fourth shell has valence shell and the number of electrons are in 1 : 1 ratio in the first, fourth and second, third shells respectively. Its corresponding oxide Y on hydrolysis gives Z. Explain the changes that take place when phenolphthalein and methyl orange are added to Z separately.

30. Calcium hydroxide on heating gives a solid product X and on treatment with hydrochloric acid gives another solid product Y. What do you observe when X and Y are exposed to atmosphere? Explain.

31. A non-metal X with electronic configuration 2, 8, 6 forms corresponding -ous and -ic acids. Identify both the acids and explain the method of dilution of the corresponding -ic acid.

32. Between H_2SO_4 and H_3PO_4, which solution possesses greater number of H+ ions in the given volume assuming that the number of molecules of both acids is equal? Give reasons.

33. Sunny and Bunny are neighbours staying in the same apartment. In summer vacation, one day they wanted to prepare some snacks for eating. During that process, the food material fell on the floor and the beautiful marble flooring became dirty. Sunny was afraid that his mother would scold him and in a hurry, brought some acid to clean the floor. But Bunny stopped him and said that it will spoil the flooring permanently. Why did Bunny say so? Give reason.

Level 3

1. Tartaric acid is one of the components in baking powder. What is its role?

2. Calcium hydroxide is added to soils before applying fertilisers. Justify the purpose.

3. What is the role of calcium hydroxide used in making mortar and white washing?

4. Why is Chile salt petre used in the manufacture of gunpowder?

5. Write formulae and names of the various salts formed when magnesium hydroxide is treated with sulphurous acid.

ASSESMENT TEST

Test 1

Directions for questions from 1 to 15

Select the correct alternative from the given choices.

1. Among the following, the strong acid is
   (a) phosphoric acid (b) carbonic acid
   (c) sulphuric acid (d) acetic acid

2. The salt formed by complete neutralisation of calcium hydroxide with sulphuric acid is
   (a) calcium sulphite (b) calcium bisulphate
   (c) calcium sulphate (d) calcium bisulphite

3. Chile salt petre and nitre are respectively
   (a) sodium nitrate and potassium nitrate
   (b) potassium nitrate and sodium nitrate
   (c) ferrous sulphate and copper sulphate
   (d) copper sulphate and ferrous sulphate

4. Identify the salt which does not contain either replaceable hydrogen ions or hydroxyl ions.
   (a) magnesium hydroxy chloride
   (b) sodium bicarbonate
   (c) disodium hydrogen phosphate
   (d) calcium carbonate

5. Monobasic weak acid among the following is
   (a) carbonic acid (b) oxalic acid
   (c) acetic acid (d) nitric acid

6. _______ on hydrolysis gives only strong acid.
   (a) sulphur dioxide (b) carbon dioxide
   (c) sulphur trioxide (d) nitrogen dioxide

7. Salt solutions are good conductors of electricity. This is due to the presence of _________ in solution state.
   (a) mobile electrons (b) mobile ions
   (c) molecules (d) both (a) and (c)

8. **Assertion (A)**: Pickles are generally not stored in tin vessels.

   Reason (R): Tin reacts with the acid components of pickles.
   (a) Both A and R are true and R is the correct explanation for A.
   (b) Both A and R are true but R is not the correct explanation for A.
   (c) A is true and R is false.
   (d) A is false and R is true.

9. Certain statements which describe different types of salts are given below:
   (1) Two basic and one acidic radicals are present in one molecule of this salt.
   (2) This salt reacts further with a base.
   (3) This salt consists of one basic radical and one acidic radical and does not react further either with an acid or with a base.
   (4) This salt is a combination of two simple salts.
   (5) This salt reacts further with an acid.

   Arrange the statements in the following order.
   (i) acidic salt (ii) basic salt
   (iii) normal salt (iv) double salt
   (v) mixed salt
   (a) 25341 (b) 25413
   (c) 25314 (d) 54321

10. Arrange the following equations in the correct sequence to obtain a normal salt.
    (1) $NaOH + Na_2HPO_4 \rightarrow Na_3PO_4 + H_2O$
    (2) $NaOH + H_3PO_4 \rightarrow NaH_2PO_4 + H_2O$
    (3) $NaOH + NaH_2PO_4 \rightarrow Na_2HPO_4 + H_2O$
    (a) 231 (b) 123
    (c) 321 (d) 312

11. **Assertion (A)**: Pure nitric acid on standing for long turns yellow.

    Reason (R): HNO_3 undergoes decomposition and forms NO_2.
    (a) Both A and R is true and R is the correct explanation for A.
    (b) Both A and R are true but R is not the correct explanation for A.
    (c) A is true and R is false.
    (d) A is false and R is true.

12. Which among the following is the application of neutralisation?
    (a) Farmers add slaked lime to soil.
    (b) People suffering from acidity are given antacid tablets.
    (c) Usage of lithium hydroxide in submarines.
    (d) All the above

PRACTICE QUESTIONS

13. Match the following

Column A	Column B
A. $CuSO_4$	( ) a. Hygroscopic
B. $Na_2CO_3.10H_2O$	( ) b. Deliquescent
C. Concentrated H_2SO_4	( ) c. Efflorescent
D. $ZnCl_2$	( ) d. Anhydrous salt

(a) A→b, B→c, C→d, D→a
(b) A→d, B→c, C→a, D→b
(c) A→c, B→d, C→b, D→a
(d) A→c, B→d, C→a, D→b

14. Role of nitre in the manufacture of gunpowder is ________.
(a) to supply oxygen
(b) to supply nitrogen
(c) to decrease the rate of combustion
(d) to absorb temperature produced by combustion

15. X is a base which is soluble in water. Then the metal present in X may be
(a) aluminium (b) copper
(c) sodium (d) manganese

Test 2

Directions questions from 1 to 15
Select the correct alternative from the given choices.

1. Which among the following is a weak acid?
(a) sulphuric acid (b) hydrochloric acid
(c) nitric acid (d) carbonic acid

2. The salt formed by the complete neutralisation of aluminium hydroxide with sulphuric acid is
(a) $Al_2(SO_3)_3$ (b) $Al_2(SO_4)_3$
(c) $Al(HSO_4)_3$ (d) $Al(HSO_3)_3$

3. Identify the common use(s) of both Chile salt petre and nitre.
(a) manufacture of gunpowder
(b) manufacture of fertiliser
(c) manufacture of glass
(d) all the above

4. Acidic and double salts among the following are respectively
(a) potash alum and sodium phosphate
(b) sodium dihydrogen phosphate and Mohr's salt
(c) Mohr's salt and calcium hydroxy chloride.
(d) sodium phosphate and Mohr's salt

5. Which of the following is a dibasic weak mineral acid?
(a) acetic acid (b) sulphuric acid
(c) phosphoric acid (d) sulphurous acid

6. If a non-metal forms only two oxides and one oxide on hydrolysis gives an acid, then the non-metal could be
(a) sulphur (b) carbon
(c) phosphorus (d) nitrogen

7. Which among the following is/are poor conductor(s) of electricity?
(a) pure H_2SO_4
(b) pure HNO_3
(c) liquefied hydrogen chloride
(d) all the above

8. **Assertion (A)**: Generally pickles are stored in glass vessels and plastic vessels.
Reason (R): The components of pickles are highly reactive towards glass/plastic.
(a) Both A and R are true and R is the correct explanation for A.
(b) Both A and R are true and R is not the correct explanation for A.
(c) A is true and R is false.
(d) Both A and R are false.

9. (1) $Ca_3(PO_4)_2$ (2) $Ca(OH)Cl$
(3) $CaOCl_2$ (4) $K_4[Fe(CN)6]$

Arrange the above salts in the order of complex salt, basic salt, normal salt and mixed salts, respectively.

(a) 4231 (b) 2431

10. Arrange the following equations in the correct sequence to obtain a normal salt.
(1) $Al(OH)Cl_2 + HCl \rightarrow AlCl_3 + H_2O$
(2) $Al(OH)_3 + HCl \rightarrow Al(OH)_2Cl + H_2O$
(3) $Al(OH)_2Cl + HCl \rightarrow Al(OH)Cl_2 + H_2O$
(a) 321 (b) 231
(c) 123 (d) 312

11. **Assertion (A)**: Pure H_2SO_4 is colourless.
Reason (R): On long standing H_2SO_4 undergoes decomposition to form SO_2.

(a) Both A and R are true and R is the correct explanation for A.

(b) Both A and R are true and R is not the correct explanation for A.

(c) A is true and R is false.

(d) Both A and R are false.

12. Which among the following is an antacid pair?

(a) $Mg(OH)_2$, NaOH (b) $Mg(OH)_2$, $Al(OH)_3$

(c) KOH, NaOH (d) $Al(OH)_3$, NaOH

13. Match the following:

Column A	Column B
A. CaO	( ) a. Deliquescent
B. $MgCl_2 . 6H_2O$	( ) b. Hydrated salt, but not deliquescent
C. $FeSO_4 . 7H_2O$	( ) c. Anhydrous and deliquescent salt
D. $MgSO_4$	( ) d. Hygroscopic

(a) A → d, B → a, C → b, D → c

(b) A → a, B → d, C → b, D → c

(c) A → b, B → c, C → d, D → a

(d) A → d, B → c, C → b, D → a

14. The gaseous products that are formed when gun-powder is subjected to heating are dissolved in water. Identify the products formed on dissolution.

(a) H_2CO_3, HNO_3 (b) H_2SO_3, H_2SO_4

(c) H_2SO_4, HNO_2 (d) H_2SO_3, H_2CO_3

15. Identify the pair of water-soluble bases:

(a) copper oxide and potassium oxide

(b) copper oxide and sodium oxide

(c) sodium, oxide and potassium oxide

(d) aluminium oxide and magnesium oxide

TEST YOUR CONCEPTS

Fill in the blanks

1. Mineral acids
2. metal
3. ammonia water
4. hydrogen
5. CO_2, H_2O
6. yellow
7. $CaSO_4 \frac{1}{2} H_2O$, calcium sulphate hemihydrate
8. Anhydrous salt
9. white
10. Copper sulphate

Multiple choice questions

11. d
12. c
13. b
14. b
15. d
16. b
17. a
18. c
19. c
20. a
21. c
22. c
23. d
24. b
25. d
26. c
27. b
28. b

Match the following

29. A → b
    B → a
    C → d
    D → c, b
30. A → d
    B → a
    C → b, e
    D → f
31. A → e
    B → c, d
    C → a
    D → b

CONCEPT APPLICATION

Level 1

True or false

1. True
2. False
3. False
4. True
5. True
6. True
7. False
8. False
9. True
10. True

Fill in the blanks

11. monobasic
12. weak acid
13. sulphurous acid
14. pink
15. alkali
16. normal salt
17. $Na_2CO_3.10H_2O$
18. potassium nitrate
19. deliquescent
20. Nitric acid

Multiple choice questions

21. d
22. c
23. c
24. b
25. a
26. c
27. b
28. a
29. a
30. d
31. d
32. b
33. b
34. d
35. a
36. c
37. c
38. a

Match the following

39. A → c
    B → a
    C → e
    D → b

40. A → c
    B → d
    C → b
    D → a

41. A → e
    B → c
    C → b
    D → a

Level 2

1. c	2. c	3. a	4. a	5. c
6. a	7. c	8. b	9. a	10. d
11. b	12. c	13. a	14. b	15. d
16. b	17. b	18. d		

ASSESMENT TEST

Test 1

1. c	2. c	3. a	4. d	5. c
6. c	7. b	8. a	9. a	10. a
11. a	12. d	13. b	14. a	15. c

Test 2

1. d	2. b	3. d	4. b	5. d
6. b	7. d	8. c	9. c	10. b
11. c	12. b	13. a	14. d	15. c

CONCEPT APPLICATION

Level 1

True or false

1. The number of hydrogen ions furnished by one molecule of an acid, on dissolving in water, is called basicity of an acid. Hydrochloric acid (HCl) is a monobasic acid.

2. The acids which ionise completely in water to furnish large concentration of H^+ ions are called strong acids. The acids in which less number of molecules ionise in water to furnish H^+ less concentration of ions are called weak acids. Carbonic acid (H_2CO_3) is a weak acid.

3. $SO_2 + H_2O \rightarrow H_2SO_3$
   Sulphurous acid

4. Acids turn methyl orange solution to pink.

5. As NH_4OH is soluble in water it is an alkali.

6. Normal salt is formed by complete replacement of the hydrogen ions of an acid with metal ions. ***Example:*** PbS

7. Washing soda is sodium carbonate decahydrate with the formula $Na_2CO_3.10H_2O$.

8. Nitre is the common name of potassium nitrate.

9. Deliquescent salts change into solution form due to the absorption of water from air. Mercuric nitrate is a deliquescent salt.

10. Chile salt petre is the common name of sodium nitrate which is used in manufacture of nitric acid.

Fill in the blanks

11. Hydrochloric acid is a monobasic acid.

12. Carbonic acid is a weak acid.

13. Sulphur dioxide on hydrolysis gives sulphurous acid.

14. Acids turn methyl orange solution to pink.

15. Ammonium hydroxide is an alkali.

16. Sodium sulphate is a normal salt.

17. Formula of washing soda is $Na_2CO_3.10H_2O$ (sodium carbonate decahydrate)

18. Nitre is the common name of potassium nitrate.

19. Mercuric nitrate is a deliquescent salt.

20. Chile salt petre is used in the manufacture of nitric acid.

Multiple choice questions

21. Stearic acid is obtained from living system, i.e., from mustard oil, therefore, it is an organic acid.

22. HNO_3 is a strong acid and $Al(OH)_3$ is a weak base.

23. H_3PO_4 on dissociation furnishes $3H^+$ ions in aqueous solution, therefore, it is a tribasic acid.

24. $Ca(OH)_2$ furnishes $2OH^-$ ions in aqueous solution, therefore, it is a diacidic base.

25. $CaCl_2$ in an example of normal salt as it is formed by complete neutralisation of an acid and a base.

26. Potassium hydroxide is an extremely corrosive alkali

27. Generally, hydrolysis of metal oxides gives bases.

28. Calcium nitrate salt is formed by treating calcium oxide solutionwith nitric acid.

29. Formula of potassium hydrogen sulphite is $KHSO_3$.

30. Ammonium carbonate is a smelling salt.

31. The acids which are obtained from the minerals present in earth are called mineral acids. A weak acid is the acid in which less than 30% of molecules ionise in water. A dibasic acid is the one which furnishes two H^+ ions by one molecule of an acid. Hence, H_2SO_3 is a dibasic weak mineral acid.

32. Carbon forms two oxides. CO_2 on hydrolysis gives carbonic acid and CO is neutral. Carbon is the non-metal.
    $CO_2 + H_2O \rightarrow H_2CO_3$

33. Antacid is the substance which is used to neutralise the acidity in stomach. Generally used antacids are magnesium hydroxide and aluminium hydroxide.

34. Acids act as good conductors of electricity when they are in ionised state. Concentrated and pure H_2SO_4 concentrated and pure HNO_3, liquid HCl do not ionise, and hence are bad conductors of electricity.

35. Calcium oxide is a hygroscopic substance which absorbs moisture without undergoing any change in its physical state.

36. Magnesium oxide and potassium oxide are soluble in water and forming alkalies namely magnesium hydroxide and potassium hydroxide, respectively.

37. (i) $4Na + O_2 \rightarrow 2Na_2O$
(ii) $Na_2O + H_2O \rightarrow 2NaOH$
(iii) $NaOH + H_2SO_4 \rightarrow NaHSO_4 + H_2O$
(iv) $NaHSO_4 + NaOH \rightarrow Na_2SO_4 + H_2O$

38. (i) Ferric hydroxide and hydrochloric acid
(ii) Barium hydroxide and sulpheric acid
(iii) Sodium hydroxide and phosphoric acid
(iv) Sodium chloride
(v) Barium sulphate
(vi) Ferric phosphate

Match the following

39. $A \rightarrow c$ Carbonic acid is used in soda water and soft drinks.
$B \rightarrow a$ Tartaric acid is a component of baking powder. Tartaricacid is used in making cakes for baking process.
$C \rightarrow e$ Sodium hydroxide is used in soap industry.
$D \rightarrow b$ Magnesium hydroxide is used as an ant-acid to neutralise the acidity.

40. $A \rightarrow c$ Formula of green vitriol is $FeSO_4.7H_2O$
$B \rightarrow d$ Formula of blue vitriol is $CuSO_4.5H_2O$
$C \rightarrow b$ Formula of Epsom salt is $MgSO_4.7H_2O$
$D \rightarrow a$ Formula of washing soda is $Na_2CO_3.10H_2O$

41. $A \rightarrow e$ Methyl orange shows pink colour in an acidic solution.
$B \rightarrow c$ Phenolphthalein is colourless in an acidic solution.
$C \rightarrow b$ Turmeric shows yellow colour in an acidic solution.
$D \rightarrow a$ Litmus when added to an acidic solution gives red colour.

Level 2

1. Sulphuric acid is a strong acid, as sulphuric acid ionises completely in water to furnish large concentration of H^+ ions.

2. The non-metal with 5 valence electrons is phosphorus and 'ic' acid of phosphorus is phosphoric acid. Metal with one electron in valence shell forms a base with the formula MOH. Since the salt can react with two molecules of base, the salt contains two replaceable hydrogen molecules. The formula of salt X is KH_2PO_4.

3. SO_3 in presence of moisture turns blue litmus to red. NO is a neutral oxide and does not respond to litmus test. MgO is basic and turns red litmus to blue.

4. X is blue vitriol ($CuSO_4$. $5H_2O$) which is blue in colour. This on heating loses water of crystallisation and forms anhydrous salt which is colourless. When water is added to anhydrous salt, it forms copper sulphate pentahydrate which is blue in colour.

$$\underset{\text{Blue}}{CuSO_4.5H_2O} \xrightarrow{\text{heating}} \underset{\text{Colourless}}{CuSO_4} \xrightarrow{+H_2O} \underset{\text{blue}}{CuSO_4.5H_2O}$$

Y is limestone, $CaCO_3$ which on heating gives CaO. It is an exothermic reaction, and hence, water starts boiling.

5. 'ic' acid of sulphur with four oxygen atoms is H_2SO_4

$$H_2SO_4 + Ca(OH)_2 \rightarrow CaSO_4 + 2H_2O$$
(calcium sulphate)

6. Nitre is KNO_3, which on heating gives KNO_2 and O_2. The liberated oxygen is the source for the combustion of carbon and sulphur.

7. The components present in pickles are inert towards glass and plastics. Hence, generally, pickles are stored in glass vessels or plastic vessels.

8. Pure HNO_3 is highly unstable which undergoes decomposition giving H_2O, O_2 and NO_2. NO_2 is a reddish brown gas, dissolves in HNO_3 and imparts yellow colour.

$$4HNO_3 \rightarrow 2H_2O + 4NO_2 + O_2$$

9. Sulphuric acid can form various salts which are soluble in water. These salts can be used as fertilisers.

10. Carbonic acid is a weak acid.

11. The acid used in soft drinks is H_2CO_3. The metallic radical in salt should be Ca^{+2}. The formula of the acidic salt should be $Ca(HCO_3)_2$.

12. Aqueous solution of sulphur trioxide is acidic in nature, and hence, it turns blue litmus to red. As aqueous solution of sodium oxide is basic in nature, it turns red litmus to blue.

13. Cobalt chloride is a hygroscopic substance, and hence, absorbs moisture and forms pink coloured hydrated cobalt chloride without undergoing any change in its state.

14. $2Al(OH)_3 + 3H_2SO_4 \rightarrow \underset{\text{Aluminium sulphate}}{Al_2(SO_4)_3} + 6H_2O$

15. The components present in gunpowder are KNO_3, C and S. KNO_3 on heating gives KNO_2 and O_2. The liberated O_2 is the source for the combustion of carbon and sulphur. The products formed are CO_2 and SO_2 which dissolve in water to form H_2CO_3 and H_2SO_3.

16. The components present in pickles (generally organic acids) are highly reactive towards tin and form harmful compounds.

17. Pure nitric acid is highly unstable, and hence, undergoes decomposition forming NO_2, H_2O and O_2. NO_2 is a reddish brown gas. Dilute solution of NO_2 appears as yellow colour.

$$4HNO_3 \rightarrow 4NO_2 + 2H_2O + O_2$$

18. Nitric acid is used in the purification of metals like gold and silver since it dissolves the impurities like various metal oxides present on the surface of gold (or) gold being insoluble in HNO_3 remains insoluble.

19. Formula of phosphorous acid H_3PO_3. H_3PO_3 on dissolution in water produces only two H^+ ions. Only these two hydrogen molecules are replaceable and the other hydrogen is non-replaceable. Hence, it is a dibasic acid.

20. In common salt, trace amounts of $MgCl_2$ and $CaCl_2$ are present as impurities which are deliquescent in nature. They absorb moisture from the surroundings and are converted into solution. As a result, common salt becomes sticky.

21. The organic acids present in pickles react with iron or tin to form harmful salts. Furthermore, they corrode the container. Glass is inert towards the organic acids, and hence, the pickle did not get spoiled.

22. Galvanisation is the process of coating zinc on the surface of iron by using electrolysis. Iron is associated with impurities like oxides and carbonates on its surface. All these impurities being basic in nature are removed by treating it with hydrochloric acid. This enables uniform and firm coating of tin on the surface of iron.

23. Nitric acid is a good oxidising agent. Gold is purified when treated with HNO_3 as it is not soluble in HNO_3 and all the impurities are oxidised by nitric acid.

24. Hydrated sodium sulphate → $Na_2SO_4.\ 10H_2O$

Hydrated copper sulphate → $CuSO_4.5H_2O$

As hydrated sodium sulphate is efflorescent in nature, it loses water of crystallisation. Blue-coloured $CoCl_2$ present in container X absorbs moisture and turns to pink coloured $CoCl_2.2H_2O$

$$\underset{\text{blue}}{CoCl_2} \xrightarrow{\text{moisture}} \underset{\text{pink}}{CoCl_2.2H_2O}$$

25. HNO_3 is highly corrosive in nature. As a result, it forms an oxide layer on Al rod which acts as a passive layer. This passive layer prevents the reaction between Al and H_2SO_4. Hence, no liberation of hydrogen gas.

26. As A turns blue litmus to red, it is acidic in nature. As B turns methyl orange to yellow, it is basic in nature. As C is neutral towards both the litmus, it is neutral in nature, i.e., a normal salt solution.

27. $H_3PO_4 + NaOH \rightarrow \underset{\text{acidic salt}}{NaH_2PO_4} + H_2O$

$H_3PO_4 + 2NaOH \rightarrow \underset{\text{acidic salt}}{Na_2HPO_4} + 2H_2O$

$H_3PO_4 + 3NaOH \rightarrow \underset{\text{neutral salt}}{Na_3PO_4} + 3H_2O$

NaH_2PO_4 → sodium dihydrogen phosphate
Na_2HPO_4 → disodium hydrogen phosphate
Na_3PO_4 → sodium phosphate

28. As, Z on treatment with oxide A of carbon, gives milky appearance to the solution by forming B, Z is Ca $(OH)_2$, A is CO_2, B is $CaCO_3$.

As Z is produced by hydrolysis of Y. Y is CaO. As Y is prepared by treating X with oxygen, X is Ca.

$$\underset{(X)}{Ca} + O_2 \rightarrow \underset{(Y)}{CaO}$$

$$\underset{Y}{CaO} + H_2O \rightarrow \underset{Z}{Ca(OH)_2}$$

$$\underset{Z}{Ca(OH)_2} + \underset{A}{CO_2} \rightarrow \underset{B}{CaCO_3} + H_2O$$

29. As the element burns with a brick red flame, it is calcium with electronic configuration 2, 8, 8, 2.

$$\underset{(X)}{Ca} + O_2 \rightarrow \underset{(Y)}{CaO}$$

$$\underset{(Y)}{CaO} + H_2O \rightarrow \underset{(Z)}{Ca(OH)_2}$$

As $Ca(OH)_2$ is basic in nature, it turns phenolphthalein to pink and methyl orange to yellow.

30. $Ca(OH)_2 \rightarrow \underset{(X)}{CaO} + H_2O$

$Ca(OH)_2 + HCl \rightarrow \underset{(Y)}{CaCl_2} + H_2O$

CaO is hygroscopic in nature, absorbs moisture and remains in solid state.

$CaCl_2$ is deliquescent in nature, absorbs moisture and is converted into solution state.

31. Non-metal with electronic configuration 2, 8, 6 is sulphur. Corresponding ous and ic acids of sulphur are H_2SO_3 and H_2SO_4, respectively.

H_2SO_4 is diluted by adding acid to water with continuous stirring because addition of water to acid is associated with liberation of large amount of heat at a time which may lead to violent and explosive reaction.

32. As H_2SO_4 is a strong acid, it produces greater number of H^+ ions, i.e., complete ionisation.

$$H_2SO_4 \rightarrow 2H^+ + SO_4^{-2}$$

As H_3PO_4 is a weak acid, it produces less number of H^+ ions, i.e., partial ionisation.

$$H_3PO_4 \rightarrow 3H^+ + PO_4^{-3}$$

33. The major component of marble is $CaCO_3$. It being basic in nature, reacts with an acid to form salt and water (neutralisation reaction). Hence, acids being reactive with marble, they cannot be used for cleaning marble flooring.

Level 3

1. During baking, baking soda present in baking powder induces bitter taste and yellow colour to breads and cakes due to the alkaline nature of its aqueous solution. To neutralise bitter taste or alkaline taste and to remove yellow colour, tartaric acid is used.

2. Generally soils are slightly acidic in nature. As calcium hydroxide is basic in nature, it neutralises the acidity of soils. Calcium hydroxide is prepared from naturally occurring limestone.

3. Mortar is a mixture of calcium hydroxide, sand and water. On exposure to atmosphere calcium hydroxide reacts with CO_2 forming $CaCO_3$ which is a hard substance, and hence, strengthens the construction. $Ca(OH)_2$ is used for white washing because of the formation of white coloured $CaCO_3$ substance.

4. Gunpowder is a heterogeneous mixture of C, S and KNO_3. On heating KNO_3 undergoes decomposition and liberates oxygen which supports combustion of carbon and sulphur. During combustion a huge amount of heat is liberated and pressure increases inside the small area. As a result, explosion takes place.

5. $Mg(OH)_2$ is a diacidic base and H_2SO_3 is a dibasic acid. Hence, there is possibility of three different salts.

$Mg(OH)_2 + H_2SO_3 \rightarrow Mg(OH)\,HSO_3 + H_2O$
magnesium hydroxy bisulphite

$Mg(OH)\,HSO_3 + H_2SO_3 \rightarrow Mg(HSO_3)_2 + H_2O$
magnesium bisulphite

$Mg(OH)_2 + H_2SO_3 \rightarrow Mg(HSO_3)_2 + H_2O$
magnesium bisulphite

$Mg(HSO_3)_2 + Mg(OH)_2 \rightarrow 2MgSO_3 + 2H_2O$
magnesium sulphite

ASSESMENT TEST

Test 1

1. Sulphuric acid is a strong acid.
2. $H_2SO_4 + Ca(OH)_2 \rightarrow CaSO_4 + 2H_2O$
   calcium sulphate
3. Chile salt petre and nitre are sodium nitrate and potassium nitrate, respectively.
4. A normal salt like calcium carbonate do not contain either replaceable hydrogen ions or hydroxyl ions.
5. Acetic acid has only one replaceable H^+ ion, and hence, it is a monobasic acid and also it is a weak acid.
6. SO_3 of hydrolysis gives H_2SO_4 which is a strong acid.
   $SO_3 + H_2O \rightarrow H_2SO_4$
7. Salt solutions are good conductor of electricity due to the formation of mobile ions.
8. Pickles are generally not stored in tin vessels as tin reacts with acids present in the pickles.
9. becda
   (i) This salt reacts further with a base.
   (ii) This salt reacts further with an acid.
   (iii) This salt consists of one basic radical and one acidic radical and does not react further either with an acid or with a base.
   (iv) This salt is a combination of two simple salts.
   (v) Two basic and one acidic radicals are present in one molecule of this salt.
10. (i) $NaOH + H_3PO_4 \rightarrow NaH_2PO_4 + H_2O$
    (ii) $NaOH + NaH_2PO_4 \rightarrow Na_2HPO_4 + H_2O$
    (iii) $NaOH + Na_2HPO_4 \rightarrow Na_3PO_4 + H_2O$
11. Pure nitric acid on standing for long turns yellow due to its decomposition and the formation of NO_2, which imparts yellow colour.
12. The process due to which an acid completely reacts with a base to form a salt and water products is called neutralisation. Slaked lime is added to soil to neutralise the acidity of the soil. Persons suffering from acidity is given antacid tablets containing $Mg(OH)_2$ to neutralise acidity. Submariners in a submarine and astronauts in spaceships use lithium hydroxide to neutralise CO_2 which is exhaled.
13. A → d $CuSO_4$ is an example of anhydrous salt.
    B → c $Na_2CO_3.10H_2O$ is an example of efflorescent substance.
    C → a Conc. H_2SO_4 is an example of hygroscopic substance.
    D → b $ZnCl_2$ is an example of deliquescent substance.
14. Nitre is KNO_3, which on heating gives KNO_2 and O_2. The liberated oxygen is the source for the combustion of carbon and sulphur.
15. As X is soluble in water, X is an alkali. Hence, the base is NaOH and it contains sodium.

Test 2

1. Carbonic acid is a weak acid.
2. $2Al(OH)_3 + 3H_2SO_4 \rightarrow Al_2(SO_4)_3 + 6H_2O$
   Aluminium sulphate
3. Both Chile salt petre and nitre are used in the manufacture of gunpowder, fertiliser and glass.
4. Acidic salt is sodium dihydrogen phosphate. Double salt is Mohr's salt.
5. The acids which are obtained from the minerals present in the earth are called mineral acids. One molecule of a dibasic acid furnishes two H^+ ions, and hence, H_2SO_3 is a dibasic mineral acid which is also a weak acid.
6. Carbon forms two oxides. CO_2 on hydrolysis gives carbonic acid and CO is neutral. Carbon is the non-metal.
   $CO_2 + H_2O \rightarrow H_2CO_3$
7. Acids act as good conductors of electricity when they are in the ionised state. Pure H_2SO_4, pure HNO_3 and liquefied HCl do not ionise, and hence are bad conductors of electricity.

8. Glass and plastic do not react with the components of pickle. Hence, pickles are stored in glass/plastic vessels.

9. dbac

(i) $K_4[Fe(CN)_6]$ (ii) $Ca(OH)Cl$
(iii) $Ca_3(PO_4)_2$ (iv) $CaOCl_2$

10. (i) $Al(OH)_3 + HCl \rightarrow Al(OH)_2Cl + H_2O$

(ii) $Al(OH)_2Cl + HCl \rightarrow Al(OH)Cl_2 + H_2O$

(iii) $Al(OH)Cl_2 + HCl \rightarrow AlCl_3 + H_2O$

11. Pure H_2SO_4 is colourless.

H_2SO_4 is a stable acid. Hence, it does not undergo decomposition on long standing.

12. Antacid is a substance which is used to neutralise the acidity in stomach. Generally, antacids used are magnesium hydroxide and aluminium hydroxide.

13. $A \rightarrow d$ CaO is an example of hygroscopic substance.

$B \rightarrow a$ $MgCl_2$ is an example of deliquescent substance.

$C \rightarrow b$ $FeSO_4.7H_2O$ is an example of hydrated salt.

$D \rightarrow c$ $MgSO_4$ is an example of anhydrous salt.

14. The components present in gunpowder are KNO_3, C and S. KNO_3 on heating gives KNO_2 and O_2. The liberated O_2 is the source for the combustion of carbon and sulphur. The products formed are CO_2 and SO_2 which dissolve in water to form H_2CO_3 and H_2SO_3, respectively.

15. Sodium oxide and potassium oxide are soluble in water and form alkalies namely sodium hydroxide and potassium hydroxide, respectively.

Chapter

4

Air and Oxygen

Remember

Before beginning this chapter, you should be able to:

- know the concept of atmosphere.
- prove presence of gases in our surrounding.
- have basic concept of states of matter.

Key Ideas

After completing this chapter, you should be able to:

- have knowledge of different layers of atmosphere and their importance.
- describe composition and characteristics of air.
- understand methods of preparation and properties of important gases like oxygen, carbon dioxide, nitrogen and hydrogen.
- study the importance of these gases in nature and their uses.
- learn the cause, effect and different measures to curb air pollution.

INTRODUCTION

The planet earth has endowed us with the basic necessities of life like air, water, food, etc. Among these, air is the most vital for life, and hence, indispensable.

In the earlier classes, we have learnt that air is a homogeneous mixture of gases. This envelope of gases which surrounds the planet earth is called **atmosphere**. Atmosphere is a cloud of gases which becomes thinner with distance from the earth and the region above the atmosphere which is absolutely free of air is called **outer space**.

The atmosphere holds the air we breathe, moisture and dust particles which are associated with natural processes like formation of clouds, precipitation, etc.

Atmosphere also protects us from harmful radiations. Hence, it is considered as a **protective cover** which is further divided into five distinct layers.

LAYERS OF ATMOSPHERE

Our atmosphere is divided into five layers on the basis of variations in chemical composition, density and temperature at different heights from the surface of the earth.

1. **Troposphere**
   (a) The layer of atmosphere which begins at the earth's surface and extends up to 20 km height is called **troposphere**.

   (b) It holds the maximum percentage of the total mass of the atmosphere.

   (c) The density of the air and temperature gradually decrease with height. As we move to the upper layers, the temperature drops from about 17°C to 51°C.The fall in temperature with increase in height at the rate of 6°C/km is called **lapse rate**.

   (d) Most of the weather phenomena occur in this layer due to rapid variations in temperature.

   (e) The height and thickness of this layer vary from the equator to the poles.

2. **Stratosphere**
   (a) The layer of atmosphere above the troposphere which extends up to 50 km above the earth's surface is called **stratosphere**.

   (b) It holds 19 per cent of the atmospheric gases.

   (c) Temperature does not alter much up to a height of about 25 km from the earth's surface. Above this, temperature increases with height due to the absorption of highly energetic UV radiations. Here the temperature rises from –60°C to –15°C. This region of increase in temperature with height is called **inversion layer**. The region that extends from 25 km to 40 km above the earth's surface is called **ozonosphere** or **ozone layer**. This layer absorbs harmful UV radiations, and thus, protects life on the earth.

   (d) Aeroplanes fly in this layer.

3. **Mesosphere**
   (a) The layer above the stratosphere which extends to about 8590 km above the earth's surface is called **mesosphere**.

   (b) Temperature decreases from about –15°C to –120°C.

   (c) Meteorites burn up in this layer.

4. **Thermosphere or Ionosphere**

(a) This layer is also called upper atmosphere which extends up to 690–950 km above the earth's surface.

(b) The density of air is lesser than the mesosphere. The molecules in this layer absorb highly energetic radiations and get ionised. Hence, this layer consists of ions and free electrons.

(c) The temperature in this layer increases with height due to the absorption of high energetic radiations. The temperature can reach a maximum of 2000°C at the top of this layer.

(d) Radio waves get reflected in this layer which, thus, helps in long-distance radio communications. However, TV waves do not get reflected. Hence, artificial satellites are used for TV transmission.

(e) The space beyond the ionosphere is called **valley of silence**. Auroras appear in this layer of atmosphere.

Auroras: The auroras (northern and southern lights) appear as luminous bands or streams, which extend to altitudes of 320 km and are visible from the surface of the earth at the higher northern or southern latitudes.

Cause of Auroras: When the high energetic electrons from the sun interact with earth's magnetic field, they cause auroras. These high energetic electrons from the sun collide with the molecules like N_2 and O_2 present in the atmosphere. The electrons of O_2 or N_2 atoms excite to higher energy levels and when they return to their lower energy levels, they emit beautiful colours of light.

5. **Exosphere**

(a) It is the outermost layer of the atmosphere which extends up to 10,000 km above the earth's surface.

(b) This layer absorbs harmful ionising radiations.

The atmosphere that is made up of different layers holds various components of air. Before and during the 15th century, air was regarded as an element. However, in 1674, Mayow predicted that air consists of two components, an active component and an inactive component. In the year 1789, Lavoisier has given the ratio of these two components, that is oxygen and nitrogen in air, as 1 : 4 by volume. Later, it was experimentally proved that O_2 and N_2 are the major components of air.

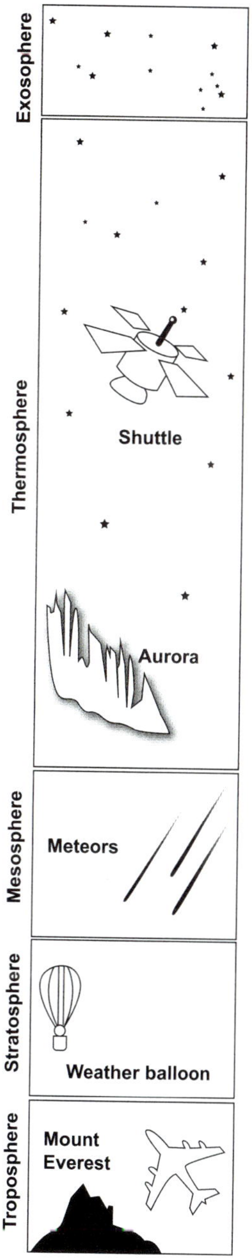

FIGURE 4.1 Layers of atmosphere

Nitrogen and oxygen are the two major components of air by volume which make up approximately 99 per cent of the dry atmosphere. The other gases found in the earth's atmosphere include water vapour, carbon dioxide, nitrous oxide, methane, ozone, hydrogen and inert gases like neon, helium, argon which are present in trace amounts.

The composition of air does not remain uniform throughout the atmosphere. It varies from place to place. Table 4.1 gives the average atmospheric composition.

TABLE 4.1 Percentage composition of air

Components of air	Percentage by volume
Nitrogen	78.084
Oxygen	20.947
Argon	0.934
Carbon dioxide	0.033
Water vapour	Variable (0 – 4)

The atmospheric layers composed of air possess weight and they exert pressure. The downward pressure exerted by air due to its weight on surroundings is called atmospheric pressure. The SI unit of atmospheric pressure is Pascal or kilo Pascal. The instrument used to measure the atmospheric pressure is called barometer. The atmospheric pressure decreases with height due to the decrease in density of air with increase in altitude. As the density of air is more in the troposphere, the pressure exerted by molecules of air is also more. Hence, the atmospheric pressure is maximum at sea level and is found to be equal to 1 atm or 760 mm of Hg.

Significance of Oxygen and Nitrogen

Nitrogen and oxygen which are major components of the atmosphere have very important associations with life. Nitrogen being an inactive component of atmosphere dilutes the activity of oxygen. Oxygen, an active component, is considered to be a supporter of life on earth as it is responsible for carrying out the important processes like respiration and combustion.

Significance of Inert Gases and Water Vapour

The significance of inert gases and water vapour is described in Table 4.2.

TABLE 4.2 Significance of other components of air

Components	Significance
Helium	To fill weather-observation balloons
Argon	To fill electric bulbs at low pressure
Neon	Used in advertisement glow tubes
Water vapour along with dust particles	Help in the formation of fog, mist, clouds, etc.

EXAMPLE

What is the role of water vapour in the formation of fog?

SOLUTION

The presence of water vapour in atmosphere influences the climatic conditions. Water vapour present in atmosphere gets condensed on dust particles during night time (winter nights) because of low temperature. Depending up on the temperature, the water vapour produces snow, fog, mist, etc. When water vapour condenses on dust particles and remains suspended in atmosphere, it is called fog.

EXAMPLE

Mention any three importances of atmosphere.

SOLUTION

(i) The air present in atmosphere is used for breathing.

(ii) The moisture and dust particles present in air help in formation of clouds.

(iii) The atmosphere protects us from harmful radiations.

EXAMPLE

Name the different layers of atmosphere.

SOLUTION

The different layers of atmosphere are

(i) Troposphere **(ii)** Stratosphere **(iii)** Mesosphere

(iv) Thermosphere **(v)** Exosphere

OXYGEN

Historical aspects: Carl Scheele, a Swedish chemist, prepared oxygen by the decomposition of nitre and mercuric oxide. This oxygen was found to support combustion, hence, he named it as **fire air** or **vital life**. Joseph Priestly, an English chemist, prepared oxygen by focusing sun rays with the help of convex lens on mercuric oxide taken in a test tube. Antoine Lavoisier proved that oxygen is an element and by assuming it as an essential constituent of an acid he named it as **oxygen** which means **acid former.**

Occurrence

It occurs both in free and combined states. It is the most abundant element in the earth's crust.

1. **Free state**: Oxygen is present in atmospheric air in free elemental state up to about 21 per cent by volume and 23 per cent by mass of the total air. Oxygen is found in dissolved state in water bodies, due to which aquatic species are able to survive.

2. **Combined state**: Oxygen is found in combined state in the most vital component of life that is water. It is also found in combined state in plants and animals. However, 50 per cent of oxygen is present in the form of minerals, such as, silicates, carbonates and oxides of various metals.

Preparation of Oxygen

I. General methods of Preparation

General methods of preparation are shown in Table 4.3.

TABLE 4.3 General methods of preparation of oxygen

Method	Chemical equation	Observation	Test
(i) Thermal decomposition of metal oxides. Mercuric oxide upon heating undergoes decomposition to produce a corresponding metal and oxygen gas. **Note**: This method is not employed for laboratory preparation because Hg vapours are poisonous and HgO is very expensive. Moreover, it produces a very small amount of O_2.	$2HgO \rightarrow 2Hg + O_2\uparrow$	Mercury vapour condenses on the cooler parts of the test tube to form a mirror-like surface.	The glowing splinter rekindles into flame.
(ii) Thermal decomposition of metal nitrates: Highly active metal nitrates on heating decompose to give metal nitrite and oxygen. **Note**: This method is not used for laboratory preparation because the rate of evolution of O_2 is slow and strong heating is required.	$2KNO_3 \rightarrow 2KNO_2 + O_2\uparrow$	Crystals of potassium nitrate melt and decompose to give a colourless gas.	The glowing splinter rekindles into flame.

II. Laboratory method of Preparation

(a) Decomposition of hydrogen peroxide

Principle: Hydrogen peroxide in the presence of a catalyst, i.e., manganese dioxide (MnO_2) undergoes decomposition to give water and oxygen at room temperature. The role of MnO_2 is to speed up the reaction and also to make the reaction take place at room temperature.

$$2H_2O_2 \xrightarrow{MnO_2} 2H_2O + O_2\uparrow$$

Note: The substance which alters the rate of reaction but does not take part in that chemical reaction is called **catalyst**.

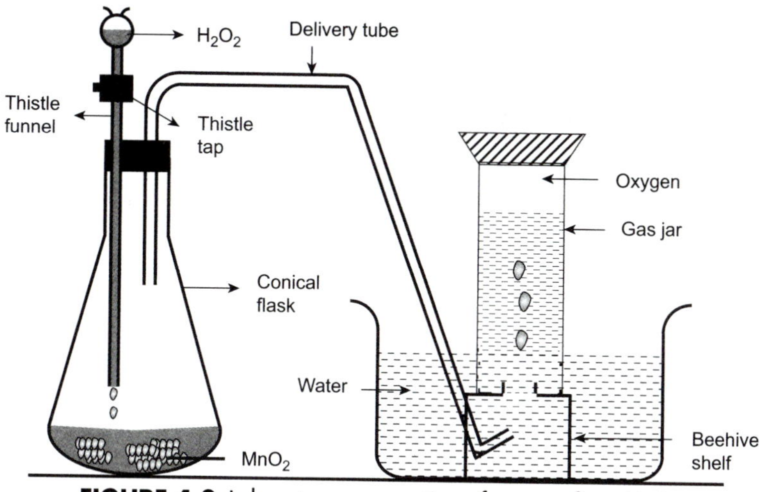

FIGURE 4.2 Laboratory preparation of oxygen from H_2O_2

Experiment: MnO_2 is taken in a clean and dry conical flask. H_2O_2 solution is added dropwise through the thistle funnel which is fitted to a conical flask with the help of a rubber stopper.

Observation: The oxygen gas evolved passes through a delivery tube and then into the beehive shelf which is placed in a trough of water. The oxygen gas, evolved, appears in the form of bubbles and is collected in the gas jar inverted over the beehive shelf. The gas rekindles a glowing splinter.

Method of collection of oxygen gas: O_2 gas evolved is collected by downward displacement of water.

Precautions

1. Heating of reaction mixture should be avoided.
2. MnO_2 used should be devoid of impurities.

Physical Properties of Oxygen

Physical properties of oxygen are shown in Table 4.4.

TABLE 4.4 Physical properties of oxygen

Physical properties	Characteristics
Colour, odour and taste	Colourless, odourless and tasteless
Solubility in water	Fairly soluble in water
Vapour density	It is slightly heavier than air
Liquefaction	Can be liquefied at high pressures and low temperatures

Note: Vapour density of a gas or vapour at a certain temperature and pressure can be defined as the ratio of mass of the certain volume of gas or vapour to the mass of the same volume of hydrogen gas at the same temperature and pressure. Hence, vapour density is a relative term.

Chemical Properties

Chemical properties of oxygen are listed in Table 4.5.

TABLE 4.5 Chemical properties of oxygen

Chemical properties	Observations and inferences	Chemical equations
I. Reaction with metals: Metals react with oxygen to form corresponding metallic oxides which are generally basic in nature.		
(i) **Reaction with sodium (Na)**: Sodium reacts with O_2 vigorously at room temperature to form sodium oxide.	Burns with a brilliant golden yellow flame.	$4Na + O_2 \rightarrow 2Na_2O$
Sodium oxide dissolves in water to form sodium hydroxide.	Sodium hydroxide turns red litmus paper to blue.	$Na_2O + H_2O \rightarrow 2NaOH$
(ii) **Reaction with magnesium (Mg)**: Magnesium reacts with oxygen upon slight heating to form magnesium oxide.	Burns with dazzling white flame	$2Mg + O_2 \rightarrow 2MgO$
Magnesium oxide dissolves sparingly in water to form magnesium hydroxide.	Magnesium hydroxide turns red litmus paper to blue.	$MgO + H_2O \rightarrow Mg(OH)_2$
II. Reaction with non-metals: Non-metals react with oxygen to form corresponding non-metallic oxides which are generally acidic in nature.		
(i) **Reaction with carbon**: When red hot charcoal or carbon is heated in an atmosphere of oxygen, it forms carbon dioxide (a colourless gas).	Sparkles and crackling sound are observed. Huge amount of heat is also liberated.	$C + O_2 \rightarrow CO_2\uparrow$
CO_2 gas dissolves in water to form carbonic acid (H_2CO_3).	H_2CO_3 turns blue litmus paper to red. Hence, CO_2 is acidic in nature.	$CO_2 + H_2O \rightarrow H_2CO_3$
Test for CO_2: When CO_2 gas is passed through lime water, a milky white precipitate of calcium carbonate ($CaCO_3$) is formed.	Lime water turns milky due to the formation of $CaCO_3$.	$Ca(OH)_2 + CO_2 \rightarrow CaCO_3\downarrow + H_2O$
(ii) **Reaction with sulphur**: Sulphur when heated in an atmosphere of oxygen burns to form sulphur dioxide.	Burns with a brilliant blue flame, SO_2 has a suffocating smell.	$S + O_2 \rightarrow SO_2\uparrow$
SO_2 when dissolved in water forms sulphurous acid (H_2SO_3)	H_2SO_3 turns blue litmus paper to red. Hence, SO_2 is acidic in nature.	$SO_2 + H_2O \rightarrow H_2SO_3$
Test for SO_2: When SO_2 gas is exposed to acidified potassium dichromate paper, the paper turns green.	Paper turns green due to the formation of $Cr_2(SO_4)_3$ chromic sulphate.	$3SO_2 + K_2Cr_2O_7 + H_2SO_4 \rightarrow K_2SO_4 + \underset{\text{green}}{Cr_2(SO_4)_3} + H_2O$
(iii) **Reaction with phosphorus**: When phosphorus is heated in an atmosphere of oxygen, it burns to produce phosphorus pentoxide.	Burns with dazzling, dense white fumes	$4P + 5O_2 \rightarrow 2P_2O_5$
P_2O_5 when dissolved in water forms phosphoric acid (H_3PO_4).	H_3PO_4 turns blue litmus paper to red. Hence, P_2O_5 is acidic in nature.	$P_2O_5 + 3H_2O \rightarrow 2H_3PO_4$

Other Chemical Properties

1. **Combustibility**: As oxygen does not burn on its own, it is a non-combustible gas. However, it is a supporter of combustion as it allows burning of other substances.

   Example: A glowing wooden splint bursts into flame in an atmosphere of oxygen.

2. **Action with litmus**: Oxygen is neither acidic nor basic in nature. It is neutral, hence, it does not respond to either blue litmus or red litmus paper.

3. **Rusting**: The process of conversion of iron into its hydrated form of oxide in a humid atmosphere is called **rusting**. The hydrated form of its oxide is called **rust**.

$$4Fe(s) + 3O_2 \rightarrow 2Fe_2O_3\ (s)$$

$$Fe_2O_3\ (s) + XH_2O\ (l) \rightarrow \underset{\text{Rust}}{Fe_2O_3.XH_2O}\ (s)$$

   (X indicates the number of water molecules which is variable)

   Iron gets corroded due to the formation of rust. The rust is a brownish flaky substance which easily crumbles from the surface of a metal. Hence, it weakens the iron structures.

   Note: Iron does not get rusted in dry atmosphere.

   Prevention: Iron is used to make different structures for varied purposes like the manufacture of machinery and tools for industrial purposes, construction of bridges, railway tracks and manufacture of furniture, bodies of cooking stoves, refrigerators, buckets, suitcases and other household goods. To prevent rusting of iron, different methods and processes are employed.

1. The metal iron is coated with other substances which prevent the exposure of iron to humid air thereby preventing rusting.

   (a) **Coating with red lead oxide or tar**: The bridges, electric poles, etc., are coated with red lead oxide paint or tar which protects iron from rusting.

   (b) **Coating with enamel paint**: Doors and windows, bodies of cars, buses and other automobiles, etc., are first given coat of red lead oxide paint followed by enamelled paint so as to enhance the appearance of the structure.

   (c) **By enamelling**: The process of coating of mixture of silicates on the surface of iron, at high temperatures is called **enamelling**. It also gives good appearance for the articles. Articles like bodies of cooking stoves, refrigerators, plates, wash basins, ovens, etc., are given this type of coating.

   (d) **Coating with oil or grease**: The movable machine or engine parts are coated with a layer of oil or grease which not only prevents rusting but also lubricates them. They cannot be coated with other substances because the coating wears off as these parts keep on moving.

2. **Galvanisation**: Iron sheets which are dipped in molten zinc are passed through heavy rollers. As a result, the iron sheets get coated with zinc metal. This process is called **galvanisation** and iron coated with zinc is called **galvanised iron**.

   Examples: Buckets, tubs, iron trunks, etc.

3. **Tinning**: The process in which a protective layer of tin is coated on iron sheets is called **tinning**. The process is similar to galvanisation.

   Example: Containers or tins used to preserve edible food materials like oil, meat, vegetables, etc.

4. **Electroplating**: The metal iron is coated with another metal by the process called **electroplating**. The metals used for coating are chromium or nickel.

   Examples: Bicycle handles, rims, bumpers of car, etc.

5. **Alloying**: The process of mixing uniformly small quantities of different metals with a particular metal in their molten state in order to improve the mechanical properties of the metal is called **alloying**. Substances like carbon and chromium are added to iron to form a homogeneous mixture called stainless steel, an **alloy** by which corrosion resistant is made.

   Percentage composition: chromium – From 12 to 20 per cent
   Carbon – From 0.07 to 0.1 per cent
   Examples: Surgical instruments, tools, utensils, cutlery, etc.

Laboratory Tests to Detect Oxygen

1. It does not respond to litmus paper as it is neutral in nature.
2. It rekindles a glowing splinter.
3. It reacts with colourless nitric oxide to form reddish brown fumes of nitrogen dioxide.
4. It turns alkaline pyrogallol solution brown.

Uses of Oxygen

1. **Respiration**: Oxygen is necessary for respiration of living beings. Oxygen inhaled during respiration is utilised in oxidising the glucose to carbon dioxide and water, which is accompanied with evolution of energy.

$$C_6H_{12}O_6 + 6O_2 \rightarrow 6CO_2 + 6H_2O + \text{energy}$$

2. **Fuels:** Oxygen is essential for burning or combustion of fuels.
3. **Artificial respiration**: It is used for artificial respiration in the places where there is no sufficient supply of oxygen. It is used by deep-sea divers, submariners, mountaineers, astronauts, aviators, etc.
4. **Medical**: A mixture of 95 per cent oxygen and 5 per cent CO_2 is called **carbogen**. It is given for the patients suffering from breathing problems.
5. **Cutting and welding**: A mixture of hydrogen and oxygen on burning gives oxyhydrogen flame of 2800°C temperature, whereas a mixture of hydrogen and acetylene produces oxyacetylene flame which has 3300°C temperature. These flames easily melt the metals, and hence, used for cutting or welding them.
6. **Space ships**: Liquid oxygen is used for burning fuel in the rockets and space ships.
7. **Explosive**: A mixture of coal, petroleum jelly and liquid oxygen is called cartridge and is used for blasting rocks.
8. **Chemical industry**: It is used in the preparation of H_2SO_4 and HNO_3.

CARBON DIOXIDE

Historical aspects: van Helmont predicted the formation of this gas when charcoal is burnt in air and he called this gas as **carbonium**. He predicted the occurrence of this gas in coal mines and found that it can extinguish candle flame. Later, Joseph Black prepared this gas by heating magnesium carbonate and he named the gas as **fixed air**.

Lavoisier named this gas as **acid carbonique** and he established it as an oxide of carbon. According to the IUPAC system, the gas was named carbon dioxide.

Occurrence: Carbon dioxide is a gas which occupies about 0.003 per cent of the total volume of air. It is the fifth abundant gas in the atmosphere.

1. **Free state**: It is found in the free state in the atmosphere and also found in water bodies like oceans, lakes, rivers, etc., in the dissolved form.
2. **Combined state**: It is generally found in combined state in the form of carbonate minerals like lime stone, magnesite, dolomite, etc.

Preparation of Carbon Dioxide

1. **General methods of preparation**

(a) Combustion: CO_2 is released into the atmosphere when carbon-containing fuels like coal, natural gas, etc., are burnt.

Examples: $C + O_2 \rightarrow CO_2\uparrow + \text{heat}$

$$CH_4 + 2O_2 \rightarrow CO_2\uparrow + 2H_2O + \text{heat}$$

(b) Decomposition of metal carbonates: Metal carbonates on strong heating decompose to produce CO_2 gas and corresponding metal oxides.

Exceptions: Sodium carbonate and potassium carbonate.

Examples: $CaCO_3 \xrightarrow{\Delta} CaO + CO_2\uparrow$

$$CuCO_3 \xrightarrow{\Delta} CuO + CO_2\uparrow$$

$$ZnCO_3 \xrightarrow{\Delta} ZnO + CO_2\uparrow$$

(c) Decomposition of metal hydrogen carbonates: Active metal hydrogen carbonate like sodium hydrogen carbonate, commonly called baking soda, on heating produces CO_2 gas along with the formation of corresponding metal carbonate and water.

$$2NaHCO_3 \rightarrow Na_2CO_3 + H_2O + CO_2\uparrow$$

Sodium bicarbonate Sodium carbonate or sodium hydrogen carbonate

(d) Action of dilute acids on metal carbonates and hydrogen carbonates. Metal carbonates and hydrogen carbonates when treated with dilute acids liberate CO_2 gas with the formation of corresponding salts.

$$Na_2CO_3 + H_2SO_4 \rightarrow Na_2SO_4 + H_2O + CO_2\uparrow$$

$$Ca(HCO_3)_2 + 2HNO_3 \rightarrow Ca(NO_3)_2 + 2H_2O + 2CO_2$$

2. **Laboratory method of preparation**

Principle: Dil HCl is made to react with marble chips which contain calcium carbonate to evolve carbon dioxide gas.

$$CaCO_3 + 2HCl \rightarrow CaCl_2 + H_2O + CO_2\uparrow$$

Process: A conical flask fitted with a two-holed rubber cork is taken. Dil HCl is added dropwise through thistle funnel which is inserted into the conical flask through one hole of rubber cork. CO_2 gas evolved passes through a delivery tube, which is inserted into the conical flask through another hole of rubber cork into the gas cylinder.

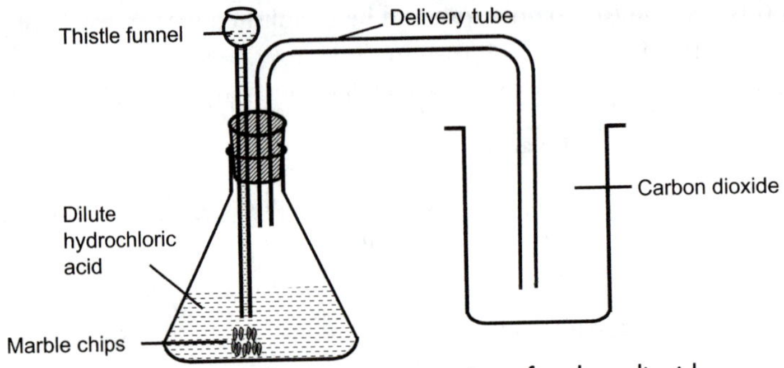

FIGURE 4.3 Laboratory preparation of carbon dioxide

Method of collection of gas: CO_2 being heavier than air is collected by upward displacement of air.

Note: Diluted H_2SO_4 is not used for the preparation of CO_2 because the calcium sulphate formed being insoluble in water forms a layer over marble chips, and thus, marble chips do not come in contact with dil H_2SO_4 further.

$$CaCO_3 + H_2SO_{4(dil)} \rightarrow CaSO_4\downarrow + H_2O + CO_2\uparrow$$

Physical Properties of CO_2

Physical properties of CO_2 are shown in Table 4.6.

TABLE 4.6 Physical Properties of carbon dioxide

Physical properties	Characteristics
Colour	Colourless
Odour	Odourless
Solubility in water	Highly soluble. Solubility increases with increase in pressure
Vapour density	Heavier than air, vapour density = 22
Nature	Acidic. Hence, changes blue litmus to red
Formation of dry ice	At very high pressures when it is allowed to expand suddenly through a small nozzle it forms solid carbon dioxide which is known as **dry ice**

Chemical Properties of CO_2

Chemical properties of CO_2 are shown in Table 4.7.

TABLE 4.7 Chemical Properties of carbon dioxide

Properties	Observation and inference
(i) **Combustibility**	It does not burn and it does not support burning of other substances in its atmosphere. Hence, it is neither combustible nor a supporter of combustion
(ii) **Reaction with water**	CO_2 reacts with water to form carbonic acid. Hence, CO_2 is acidic in nature. $CO_2 + H_2O \rightarrow H_2CO_3$
(iii) **Acidic properties of CO_2** (a) Reaction with alkali	It reacts with alkali (base) to form salt and water. This reaction is called **neutralisation**. ***Examples:*** $CO_2 + 2NaOH \rightarrow Na_2CO_3 + H_2O$

Properties	Observation and inference
(b) Reaction with lime water	CO_2 reacts with lime water to form a milky white precipitate, i.e., $CaCO_3$. $Ca(OH)_2 + CO_2 \rightarrow CaCO_3$ **Note**: When excess CO_2 is passed through the above solution, the precipitate dissolves due to the formation of $Ca(HCO_3)_2$ $CaCO_3 + H_2O + CO_2 \rightarrow C(HCO_3)_2$
(c) Reaction with metals	Metals like Na, K and Ca react with CO_2 to form corresponding metal carbonates while Mg and Al give corresponding oxides. $4Na + 3CO_2 \rightarrow 2Na_2CO_3 + C$ $2Mg + CO_2 \rightarrow 2MgO + C$
(iv) Reaction with non-metals	CO_2 when reacts with red hot coke gets reduced to CO, a very poisonous gas. $\underset{\text{Red hot coke}}{C} + CO_2 \xrightarrow{\Delta} 2CO$

Uses of Carbon Dioxide

CO_2 finds many applications in various fields like industry, medicine, agriculture, etc. It is a natural raw material available for the plants to synthesise the food material by the process called photosynthesis. Moreover, it maintains the temperature of the earth. The various other uses of CO_2 are given below.

1. **Manufacture of useful chemicals**: It is extensively used in the production of industrial chemicals such as washing soda, baking soda ($NaHCO_3$), white lead (Pb $(OH)_2.PbCO_3$), etc.
2. **Soft drinks**: It is dissolved in soft drinks under high pressure so as to impart tangy taste.
3. **Medical purpose**: Carbogen, which consists of 5 per cent CO_2 and 95 per cent O_2 stimulates breathing (artificial respiration), and hence, it is used by patients suffering from asthma, pneumonia, etc.
4. **Refrigeration**: Dry ice is used as a refrigerant to preserve perishable foods like meat, fish, fruits, vegetables, etc., for longer time.
5. **Preservation of foodstuffs**: It is used to preserve food grains.
6. **Regeneration of oxygen**: CO_2 present in the atmosphere is used by the plants to prepare their food by the process called **photosynthesis**. In this process, CO_2 and H_2O combine in the presence of sunlight to form glucose and O_2 which is added to the atmosphere.

$$6CO_2 + 6H_2O \xrightarrow{\text{(Sunlight)}} \underset{\text{(Glucose)}}{C_6H_{12}O_6} + 6O_2 \uparrow$$

7. **Fire extinguisher**: CO_2 is a non-supporter of combustion, non-combustible and being heavier than air it can be used to put off fires. The device used to put off fires is called **fire extinguisher.**

Laboratory Tests for Carbon Dioxide

1. When a lightened candle is exposed to the atmosphere of CO_2, it gets extinguished.
2. It turns lime water milky and when excess CO_2 is passed through the solution, precipitate formed disappears.

NITROGEN

Occurrence

1. Nitrogen exists in a free state in the atmosphere. It is the major component of air which is present up to about 78 per cent by volume.
2. It is also found in a combined state in the form of salts like KNO_3, $NaNO_3$ and also in organic matter such as proteins, etc.

Uses

1. It dilutes the activity of oxygen, and thus, prevents rapid burning.
2. It is used in the manufacture of ammonia. Compounds of ammonia are used as fertilisers.
3. Nitrogen is required for plants and animals to synthesise proteins. The free nitrogen cannot be assimilated by plants. Hence, nitrogen present in the atmosphere is converted into compounds by which it can be assimilated by plants. The process of conversion of atmospheric nitrogen into nitrates is called **nitrogen fixation**.

   During lightening, N_2 combines with O_2 to form NO which further reacts with O_2 to form NO_2.This gas dissolves in the water droplets of clouds and form HNO_3. During precipitation, it reaches soil and reacts with metal carbonates to form corresponding nitrates, and these nitrates are assimilated by plants.

   $$N_2 + O_2 \rightarrow 2NO$$

   $$2NO + O_2 \rightarrow 2NO_2$$

   $$4NO_2 + 2H_2O + O_2 \rightarrow 4HNO_3$$

   $$4HNO_3 + MCO_3 \rightarrow M(NO_3)_2 + CO_2 + H_2O$$

   Leguminous plants contain symbiotic bacteria in their root nodules which absorb atmospheric nitrogen and convert nitrogen into various compounds of nitrogen. The plants can easily assimilate and also the soil becomes enriched with nitrogen compounds.
4. Nitrogen being chemically less reactive (inert in nature), is used in the preservation of foods.

HYDROGEN

Hydrogen is the lightest among all elements. Atmospheric air is found to contain only trace amounts of hydrogen. Hydrogen gas was first prepared by the action of H_2SO_4 on iron. In 1762, Henry Cavendish for the first time established the elemental nature of hydrogen. He prepared this gas by the action of H_2SO_4 on zinc. Lavoisier named it hydrogen which means water-former (Hydro-water, gen-producing).

Occurrence

Availability of hydrogen gas in a free state is very rare. Although it is present in trace amounts in the atmosphere, hydrogen is present in the sun's atmosphere up to 1.1 per cent. Very minute quantities of hydrogen are present in volcanic gases.

Most of the hydrogen is present in a combined state in nature. Hydrogen is present in various carbon compounds in the form of proteins, nucleic acids, vitamins, hormones, enzymes, etc., in plants and animals. Most of the naturally occurring substances such as wood, cotton, paper pulp, rubber, etc., contain hydrogen. Water is the most important source of hydrogen as hydrogen is present up to 11.1 per cent by weight in water.

Preparation of Hydrogen

(i) General methods of preparation

General methods of preparation of hydrogen are shown in Table 4.8.

TABLE 4.8 General methods of Preparation of hydrogen

Method	Equation	Observation
Reaction of active metals with water	$2Na + 2H_2O \rightarrow 2NaOH + H_2\uparrow$	Na and K $\rightarrow$ Reaction is violent. They float on the surface of water. Ca $\rightarrow$ Brisk reaction and metal sinks in water.
Action of steam on hot metals	$Mg + H_2O \rightarrow MgO + H_2\uparrow$ $Fe + H2O \rightleftharpoons Fe_3O_4 + H_2\uparrow$	Reaction is fast for Mg and slow for Al, Zn and Fe.
Action of dil acids on metals	$Zn + 2HCl \rightarrow ZnCl_2 + H_2\uparrow$	—
Action of concentrated alkali on metals	$2NaOH + 2Al + 2H_2O \rightarrow$ $2NaAlO_2 + 3H_2\uparrow$	Reaction takes place only on boiling.

(ii) Laboratory method of preparation

Principle: All the metals present above hydrogen in the reactivity series on reaction with dil HCl or dil H_2SO_4 liberate hydrogen gas. Zinc metal is preferred for this reaction.

$$Zn + 2HCl \rightarrow ZnCl_2 + H_2\uparrow$$

$$Zn + H_2SO_4 \rightarrow ZnSO_4 + H_2\uparrow$$

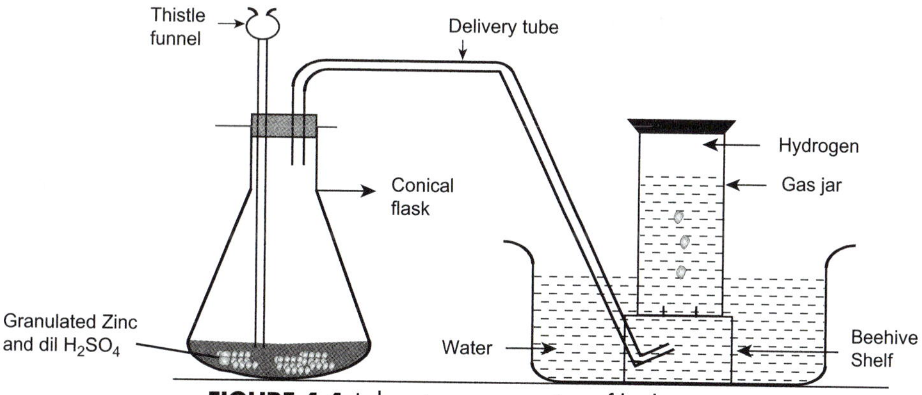

FIGURE 4.4 Laboratory preparation of hydrogen

Experiment: Some granulated zinc pieces are taken in a flat-bottom flask fitted with a two-holed air-tight cork. In one hole, a thistle funnel is inserted and in another hole, a long delivery tube is inserted. Dil H_2SO_4 is added to the flask dropwise through the thistle funnel.

Observation: Effervescence is observed during the initial stages of the reaction. Then, a colourless and odourless gas evolves.

Collection of gas: Hydrogen is collected by the downward displacement of water.

Purification of hydrogen gas: Since impure zinc is used for hydrogen gas preparation, the hydrogen gas collected is found to contain some impurities such as H_2S, SO_2, PH_3, AsH_3 and CO_2. These impurities are removed in stages.

TABLE 4.9 Removal of impurities associated with hydrogen gas

Impurities	Substances used for removal
(a) AsH_3, PH_3	Silver nitrate solution
(b) H_2S	Lead nitrate solution
(c) CO_2, SO_2, NO_2	Caustic potash
(d) H_2O (water vapour)	$CaCl_2$ or P_2O_5 (drying agent)

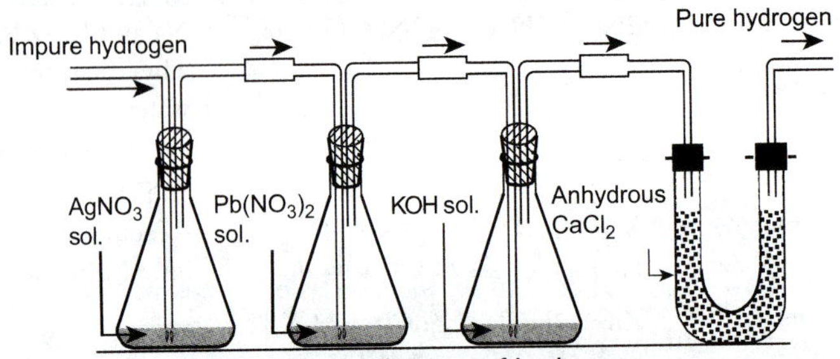

Figure 4.5 Purification of hydrogen gas

Collection of purified hydrogen gas: To obtain pure and dry hydrogen, pure hydrogen gas is collected over mercury.

Precautions

1. The entire apparatus should be kept air tight.
2. Contact with flame should be avoided.
3. Collection of the gas should be done only after the expulsion of air.
4. The end of the thistle funnel should almost touch the bottom.

Although all active metals liberate hydrogen gas, zinc is the preferred metal and the other metals are not used for this preparation method.

TABLE 4.10 Reasons for not using other metals for the preparation of hydrogen gas

Metals	Reason for not using
Sodium and potassium	Violent reaction with explosion
Calcium and magnesium	Expensive metals
Aluminium	Formation of protective coating of Al_2O_3 that renders the metal passive
Iron	Reversible reaction
Lead	Formation of insoluble $PbCl_2$ or $PbSO_4$ stops the further reaction

(iii) Industrial methods of preparation

1. **Electrolysis of water**: Electrolysis of acidulated water liberates oxygen at the anode and hydrogen at the cathode. In fact, this method is employed for the preparation of oxygen gas where hydrogen is obtained as a by-product.

$$2H_2O \rightarrow 2H_2 + O_2$$

2. **Bosch process**

Principle: Water gas is a mixture of carbon monoxide and hydrogen. Water gas can be prepared by passing steam over red hot coke.

$$C + \underset{\text{steam}}{H_2O} \rightarrow \underset{\text{water gas}}{CO + H_2}$$

Catalytic oxidation of water gas gives hydrogen gas.

$$CO + H_2 + H_2O \xrightarrow[450°C-500°C]{Fe_2O_3} CO_2 + 2H_2$$

Process

(a) Superheated steam at 170°C, when passed through red hot coke or charcoal, gives a mixture of CO gas and hydrogen known as water gas.

(b) Water gas obtained is made to pass over Fe_2O_3 or Cr_2O_3 along with superheated steam at 500°C. Carbon monoxide present in water gas gets oxidised to CO_2 leaving hydrogen gas.

(c) The CO_2 gas is removed by compressing the gas at 30 atm pressure and passing it through water. Hydrogen being insoluble remains as such.

$$H_2O_{(1)} + CO_{2(g)} \xrightarrow{30\ atm} H_2CO_3$$

(d) The water vapour present in the gas can be removed by cooling the gas to –20°C.

(e) The traces of carbon monoxide remaining unoxidised can be removed by passing the gas through ammonical cuprous chloride solution.

$$CuCl + CO + 2H_2O \rightarrow \underset{\text{(Addition compound)}}{CuCl.\ CO.\ 2H_2O}$$

Bosch process can also be carried out for the preparation of hydrogen from natural gas. Natural gas contains methane as the major constituent. Methane on mixing with superheated steam at high temperature followed by compression to 30 atm pressure gives water gas. This is subjected to Bosch process for getting pure hydrogen gas.

Physical Properties of Hydrogen

Physical properties of hydrogen are discussed in Table 4.11.

TABLE 4.11 Physical properties of hydrogen

Physical properties	Description
Odour	No characteristic odour for pure hydrogen. However, fishy odour of hydrogen gas is attributed to the presence of impurities
Solubility	Negligible
Density	Least density
Action on litmus	Neutral
Liquefaction	Difficult due to very low critical temperature and not possible at room temperature
Diffusion	Rate of diffusion is very high

Adsorption or Occlusion

In addition to the normal physical properties, hydrogen shows a specific property called adsorption or occlusion. The accumulation of molecules of one compound over the surface of another substance is called **adsorption**. Hydrogen gas gets adsorbed on the surface of some metals like palladium, platinum, nickel, etc. This phenomenon is called occlusion.

Chemical Properties of Hydrogen

1. **Reaction with air**: Hydrogen burns in air, i.e., it reacts with oxygen of air forming water. A blue flame is also observed.

$$2H_2 + O_2 \rightarrow 2H_2O$$

2. **Reaction with non-metals**: Hydrogen reacts with most of the non-metals forming corresponding hydrides.

   (a) Reaction with chlorine: Hydrogen reacts with chlorine in the presence of sunlight to give hydrogen chloride. This is a photochemical reaction.

$$H_2 + C_2 \xrightarrow{\text{light}} 2HCl$$

   (b) Reaction with nitrogen: Nitrogen and hydrogen react in 1 : 3 ratio at high temperature and high pressure to form ammonia.

$$N_2 + 3H_2 \rightleftharpoons 2NH_3 + \text{heat}$$

   (c) Reaction with sulphur: On reaction with molten sulphur, hydrogen gas gives hydrogen sulphide.

$$H_2 + S \rightarrow H_2S$$

3. **Reaction with metals**: Hydrogen reacts with highly electropositive metals to form metallic hydrides.

Examples: $2K + H_2 \xrightarrow{\Delta} 2KH$

$$2Na + H_2 \xrightarrow{\Delta} 2NaH$$

$$Ca + H_2 \longrightarrow CaH_2$$

4. **Reaction with metal oxides**: Oxides of less reactive metals react with hydrogen to form the corresponding metal and water. This is due to the reduction of metallic oxide.

Examples:

$$\underset{\text{white}}{ZnO} + H_2 \rightarrow \underset{\text{bluish white}}{Zn} + H_2O$$

$$\underset{\text{brown}}{2Fe_2O_3} + 6H_2 \rightarrow \underset{\text{grey}}{4Fe} + 6H_2O$$

$$\underset{\text{yellow}}{PbO} + H_2 \rightarrow \underset{\text{grey}}{Pb} + H_2O$$

$$\underset{\text{black}}{CuO} + H_2 \rightarrow \underset{\text{red}}{Cu} + H_2O$$

5. **Hydrogenation**: Catalytic addition of hydrogen to some carbon compounds is called **hydrogenation**.

$$\underset{\text{Ethylene}}{H_2C = CH_2} + H_2 \rightarrow \underset{\text{Ethane}}{H_3C - CH_3}$$

This principle is used in the process of hydrogenation of oils or hardening of oils.

Hydrogenation of oils: Vegetable oil, which contains carbon and oxygen in greater percentage and contains less amount of hydrogen, cannot be solidified easily due to less percentage of hydrogen.

Examples: Groundnut oil and sunflower oil

Animal oils, such as ghee, which contain hydrogen in greater percentage get solidified easily under normal conditions. Vegetable oils also get solidified when hydrogen gas is passed through them at high pressure and high temperature in the presence of nickel catalyst.

Examples: Palm oil, vanaspathi and soap

Uses of Hydrogen

1. In the synthesis of industrial chemicals
   (a) Manufacture of ammonia (Haber's process)

$$N_2 + 3H_2 \underset{}{\overset{Fe}{\rightleftharpoons}} 2NH_3 + \text{heat}$$

   (b) Manufacture of hydrogen chloride

$$H_2 + Cl_2 \xrightarrow{\text{sunlight}} 2HCl$$

   (c) Manufacture of methyl alcohol

$$[CO + H_2] + H_2 \xrightarrow[\text{Catalyst}]{\Delta} CH_3OH$$

   (d) Manufacture of dalda/vanaspathi and soap by the process of hydrogenation
   (e) Manufacture of petrol/gasoline from coal

2. (a) **Fuel**: Many fuel gases contain hydrogen as a major constituent. This is because of high heat of combustion of hydrogen.

   Examples: Water gas, coal gas, semi-water gas, etc.

   (b) Electrical cells called fuel cells are also designed in which hydrogen gas is used.
   (c) **Oxyhydrogen blow torch**: A mixture of hydrogen and oxygen is burnt to give a high temperature of 2800°C. This is used for welding of metals and for melting platinum and quartz.
   (d) Liquid hydrogen is used as a rocket fuel.
   (e) **Atomic hydrogen torch**: When an electric arc is generated in a stream of hydrogen between tungsten electrodes, the hydrogen molecules dissociate to give hydrogen atoms by absorbing large amount of energy. The atoms recombine to form hydrogen molecules by evolving the absorbed heat energy. As a result, high temperature is generated. This can be made use of in welding metals and alloys.

3. **Metallurgy**: Hydrogen is used as a good reducing agent in the extraction of some metals from their oxides.

   Examples: Tungsten and molybdenum

$$WO_3 + 3H_2 \rightarrow W + 3H_2O$$

Nascent hydrogen: It is also called newly born hydrogen. It is denoted by [H]. Nascent hydrogen can reduce compounds that do not readily react with normal hydrogen. Nascent hydrogen is usually produced by the reaction of metals with acids, e.g., zinc and sulphuric acid.

Applications of Nascent Hydrogen

1. Nascent hydrogen can be used as a rocket propellant. It could be stored in liquid helium to prevent it from recombining into molecular hydrogen.
2. Nascent hydrogen is claimed to reduce nitrites to ammonia or arsenic to arsine even under normal conditions.
3. It has temporary bleaching action.

AIR POLLUTION

The composition of atmospheric gases changes due to the addition of other gases and suspended particles mainly by human-induced activities. This built up of other gases and suspended particulate matter is toxic to organisms living on the earth's surface. The variation in the composition of atmospheric air due to the release of harmful gases is called **air pollution**.

A substance when added to atmosphere producing adverse effects to human kind and surroundings is called **air pollutant**.

Examples: CO, chlorofluorocarbons (CFCs), O_3 in troposphere, etc.

Major Factors That Contribute to Air Pollution

1. Burning of fossil fuels
2. Deforestation
3. Volcanic eruption
4. Emissions from industries and vehicles
5. Agricultural activities
6. Lack of civic sense by the people
7. Pollination
8. Forest fires

Among the above-mentioned factors volcanic eruptions, forest fires and pollination are natural ways of contributing to air pollution.

TABLE 4.12 Pollutants that are released into the atmosphere due to various processes and their effects on environment

Pollutants	Sources	Processes	Effect on environment
Carbon monoxide	Automobiles	Incomplete combustion of fuels	Respiratory problems (reduction in intake of oxygen)
Sulphur dioxide	(i) Automobiles (ii) Industries (iii) Thermal power plants	(i) Incomplete combustion of fuels (ii) Metals extraction (iii) Paper making	(i) Causes acid rain (ii) Respiratory disorders
Nitrogen oxides	(i) Automobiles (ii) Industries	Incomplete combustion of fuels	(i) Formation of smog (ii) Causes acid rain
Carbon dioxide	(i) Automobiles (ii) High density of population (iii) Decrease in forest area	Combustion of fuels and deforestation	Global warming
Ozone	Industries		Global warming
CFCs	(i) Air conditioners, organic chemicals (ii) Refrigeration systems		Depletion of ozone layer
Metals	(i) Automobiles (Pb) (ii) Thermal power plants (Pb, Hg, As, Co, Cu)	(i) Present in petrol as anti-knock agents (ii) Chemical reactions	(i) Nervous disorders (ii) Problems of digestion (iii) Cancer
Hydrocarbons	(i) Automobiles (ii) Degradation of organic wastes	Exhaust gas	(i) Carcinogenic
Fly ash	Thermal power plants	Combustion of coal	(i) Metal oxides and toxic metals of fly ash cause health problems (ii) Silica causes bronchitis and lung cancer (iii) Contamination of ground water and clogging of drainage
Suspended particulate matter (SPM)	(i) Automobiles (ii) Industries (iii) Formation of smog	Exhaust gas	(i) Lung and respiratory disorders

Hazards of Air Pollution and Their Prevention

1. **Acid rain**: The gases such as NO_2 and SO_2 that are released from factory chimneys, burning of fossil fuels, eruption of volcanoes, etc., dissolve in rain water forming HNO_3 and H_2SO_4, respectively. This is called **acid rain**.

$$2SO_2 + O_2 \rightarrow 2SO_3$$

$$SO_3 + H_2O \rightarrow H_2SO_4$$

$$2NO_2 + H_2O \rightarrow HNO_3 + HNO_2$$

Effect:

(a) The fertility of the soil is lost or reduced.

(b) Damage to the crops.

(c) Pollutes water bodies like river, seas, etc.

(d) Corrodes the marble statues or monuments or buildings.

Prevention:

(a) Minimise the usage of fossil fuels. Fossil fuels such as petrol, diesel are replaced by CNG in some metropolitan cities.

(b) Use non-conventional sources of energy.

(c) Dispose industrial wastes safely.

2. **Greenhouse effect and global warming**: The solar radiation is absorbed by the surface of oceans and vegetation on land. This absorbed radiation is transformed into heat and is re-radiated back into the atmosphere. Certain gaseous molecules present in the atmosphere absorb the infrared radiation (IR) and re-radiate the heat in all directions. Hence, these gases maintain the temperature of the earth's surface. The gases which absorb these IR are called **greenhouse gases** and the process is called **greenhouse effect**. This process supports life on earth. The greenhouse gases are CO_2, H_2O, N_2O, CH_4, O_3, CFCs, etc.

Increase in the levels of these gases results in gradual increase of temperature of earth's surface. This increased effect caused due to increase in the air pollutants is called **global warming**.

Effects of global warming:

(a) Melting of ice caps and glaciers.

(b) Increase in frequency of floods, soil erosion unseasonal rains, etc.

Prevention:

(a) Reduction in the use of fossil fuels.

(b) Controlling deforestation.

(c) Restricting the use of CFCs.

3. **Ozone depletion**: Ozone present in the troposphere contributes to the greenhouse effect. Ozone present in the stratosphere plays a vital role in protecting the life on earth's surface from harmful UV radiations. Ozone layer acts as a shield as it absorbs highly energetic harmful UV radiations.

Pollutants, such as CFCs, upon reaching the stratosphere react with ozone, and thus, deplete the ozone layer in a period of time. Hence, UV radiations reach the tropospheric layer due to the holes created in ozone and affect plant and animal life.

Prevention: Discovery of alternative substance for CFCs used in refrigeration and aerosols.

4. **Hazards caused by carbon monoxide**: As the percentage of CO increases in the atmosphere, it affects the respiratory system. CO has 200-times more affinity for haemoglobin over oxygen and forms a stable compound called carboxy haemoglobin which causes suffocation and sometimes it is even fatal to human life.

   Prevention: Reduction in the burning of fossil fuels, cigarettes, etc.

   As the pollutants increase in atmospheric air, they affect both plant and animal life, cause respiratory problems, result in damage to monuments and so on. Particularly, the lead particles from smoke of motor engines cause damage to the nervous system.

Ways of Minimising Air Pollution

The following safe measures can be adopted so as to minimise air pollution.

1. Safe disposal of waste products.
2. Equipment like scrubbers and precipitators are to be used in industries to prevent the emissions of harmful gases.
3. Synthesis of ecofriendly chemicals.
4. Reducing the growth of population.
5. Judicial use of natural resources.
6. Restricting the usage of plastic.
7. Arresting deforestation and increasing planting or greenery.

EXAMPLE

Write a short note on acid rain.

SOLUTION

The gases such as NO_2 and SO_2 are released in the atmosphere through factory chimneys. They are also generated during the burning of fossil fuels, eruption of volcanoes, etc. These gases dissolve in rain water forming HNO_3 and H_2SO_4, respectively and cause acid rain.

$2SO_2 + O_2 \rightarrow 2SO_3$

$SO_3 + H_2O \rightarrow H_2SO_4$

$2NO_2 + H_2O \rightarrow HNO_3 + HNO_2$

EXAMPLE

Explain ozone depletion. How can ozone depletion be prevented?

SOLUTION

Ozone present in stratosphere plays a vital role in protecting the life on earth's surface from harmful UV radiations. Ozone layer acts as a shield as it absorbs highly energetic harmful UV radiations. Pollutants such as CFCs when reach stratosphere react with ozone, and thus, depletes the ozone layer in a period of time. Hence, UV radiations reach tropospheric layer and cause many undesirable effects like skin cancer, cataract, damage to plants and crops, etc.

Discovery of alternative substances for CFCs used in refrigeration systems and aerosols can prevent depletion of ozone.

EXAMPLE

A combines with B to form C and D. A and D are major contributers for global warming and both are carbon compounds. B is a supporter of combustion. Magnesium burns in the presence of B with a dazzling white flame and forms a product X which is basic in nature. With C, X forms a compound which turns red litmus paper blue. Give the respective chemical reactions involved.

SOLUTION

A is methane, B is oxygen, C is water vapour and D is carbon dioxide. Methane and CO_2 contribute to greenhouse effect and both are carbon compounds. Oxygen combines with magnesium to produce magnesium oxide with dazzling white flame.

$2Mg + O_2 \rightarrow 2MgO$, MgO is basic in nature.

$MgO + H_2O \rightarrow Mg(OH)_2$.

Magnesium hydroxide is the base which turns red litmus blue.

POINTS TO REMEMBER

- Most of the weather phenomena like the formation of clouds, rain, etc., take place in the troposphere, while the stratosphere protects the earth from harmful UV radiation and also aeroplanes fly in this region.
- The density of air is maximum near the earth's surface, i.e., in the troposphere.
- Nitrogen is an inactive component while oxygen is an active component. Nitrogen is used to preserve food materials while oxygen is required for burning and respiration.
- Nitrogen dilutes the activity of oxygen, and its compounds are used as fertilisers which are helpful for the growth of plants.
- Oxygen is prepared in the laboratory by the decomposition of H_2O_2 while CO_2 is prepared by the reaction of dil acid on marble chips. The O_2 gas obtained is collected by the downward displacement of water while CO_2 is collected by the upward displacement of air.
- CO_2 is essential for photosynthesis. Being a non-supporter of combustion, non-combustible and heavier than air, it can be used to extinguish fires. It is a greenhouse gas but leads to global warming if the amounts of CO_2 increase due to human-induced activities such as deforestation, excess usage of petrol, diesel, etc.
- Water vapour along with dust particles is responsible for the formation of clouds, fog, mist, etc.
- Rust is formed when iron articles get exposed to humid air. The formula of rust is Fe_2O_3 X H_2O. The formation of rust can be prevented by galvanisation, tinning, alloying, coating with enamel paint, etc.
- NO_2 particles, SO_2, excess CO_2, N_2O, CO, excess dust and other impurities are major pollutants in air which cause hazards like acid rain, global warming, ozone depletion, respiratory problems, etc.
- Afforestation, using CNG as a substitute for petrol and diesel, installation of catalytic converters in automobiles, electric precipitators and scrubbers in industries, etc., minimise pollution.

TEST YOUR CONCEPTS

Very Short Answer Type Questions

Directions for questions from 1 to 10

Fill in the blanks.

1. Ultraviolet rays are absorbed by ______ of atmosphere.
2. Auroras appear in the layer of atmosphere which is known as ______.
3. ________ depletes the ozone layer.
4. The ______ alters the rate of reaction.
5. In the laboratory preparation of oxygen from hydrogen peroxide, _______ is used as a catalyst.
6. _____ gas is evolved when metal carbonates are treated with dil acids.
7. The solid carbon dioxide is commonly called _______.
8. _______, an inert gas is used to fill weather observation balloons.
9. The chemical formula of rust is ________.
10. The trapping of solar heat energy by certain gaseous molecules present in the earth's atmosphere is called _______.

Directions for questions from 11 to 26

For each of the questions, four choices have been provided. Select the correct alternative.

11. With the increase in altitude, the atmospheric pressure
    (a) increases
    (b) decreases
    (c) remains the same
    (d) first increases then decreases
12. In ______, the temperature decreases with the increase in altitude.
    (a) ionosphere (b) troposphere
    (c) ozonosphere (d) stratosphere
13. Which among the following gases present in atmosphere does not contribute to global warming?
    (a) nitrogen oxides like N_2O
    (b) carbon dioxide
    (c) sulphur dioxide
    (d) water vapour
14. _______ is the most abundant element in earth's crust.
    (a) Si (b) Al
    (c) O_2 (d) Fe
15. ______ is the process of mixing of a metal with metal or non-metal to form a homogeneous mixture.
    (a) alloying (b) electroplating
    (c) tinning (d) galvanisation
16. Which compound among the following is used for the laboratory preparation of oxygen?
    (a) KNO_3 (b) $Pb(NO_3)_2$
    (c) H_2O_2 (d) HgO
17. ________ is used as a supporter of combustion in rockets and space ships.
    (a) liquid hydrogen (b) liquid oxygen
    (c) coke (d) liquid carbon dioxide
18. A burning splinter gets extinguished in an atmosphere of CO_2. Identify the substance which continues to burn in an atmosphere of CO_2.
    (a) coke (b) magnesium
    (c) sulphur (d) phosphorus
19. Soda water contains excess _______in dissolved state.
    (a) H_2 (b) CO_2
    (c) N_2 (d) O_2
20. ________ is used in the refrigeration of sea food.
    (a) chlorofluoro carbons (b) dry ice
    (c) carbon dioxide gas (d) inert gases
21. Excess intake of O_2 by a person results in several ill-effects. Which of the following activities leads to excess intake of O_2?
    (a) deep-sea diving
    (b) mountaineering
    (c) travelling in an aeroplane
    (d) walking on the moon
22. Identify the component which when present in the troposphere is a pollutant and when present in upper layers of atmosphere is a protector of life.
    (a) chlorofluro carbons (b) ozone
    (c) nitrous oxide (d) water vapour

PRACTICE QUESTIONS

23. O_2 and CO_2 prepared in the laboratory can be collected by ________.
(a) downward displacement of water and downward displacement of air, respectively
(b) downward displacement of water and upward displacement of air, respectively
(c) upward displacement of water in both the cases
(d) downward displacement of water in both the cases

24. Which of the following process is not associated with the formation of metal nitrates in soil?
(a) acid rain
(b) nitrogen fixation by leguminous plants
(c) lightening followed by precipitation
(d) formation of auroras

25. Arrange the following layers of the atmosphere in the decreasing order of their altitudes.
(1) stratosphere (2) troposphere
(3) mesophere (4) exosphere
(5) ionosphere
(a) 34512 (b) 21354
(c) 42531 (d) 45312

26. Arrange the components of air in the increasing order of their percentage by volume.
(1) nitrogen
(2) carbon dioxide
(3) oxygen
(4) argon
(a) 2431 (b) 1342
(c) 4321 (d) 2134

Directions for questions from 27 to 29

Match the entries given in Column A with the appropriate ones in Column B. More than one suitable content of Column B can also be considered correct for an entry in Column A and some content given in Column B may not be the appropriate one for any entry in Column A.

27.

Column A		Column B
A. Carbon dioxide	( )	a. Acid rain
B. Sulphur dioxide	( )	b. Cutting and welding
C. H_2 and O_2	( )	c. Weather observation balloons
D. Helium	( )	d. Global warming
	( )	e. Aerated drinks

28.

Column A		Column B
A. Zinc	( )	a. Bridges
B. Silicates	( )	b. Galvanisation
C. Tar	( )	c. Surgical instruments
D. Red lead oxide	( )	d. Enamelling
	( )	e. Electric poles

29.

Column A		Column B
Reactants		**Gaseous products**
A. Hydrogen peroxide	( )	a. Nitrogen
B. Calcium carbonate	( )	b. Oxygen
C. Zinc granules	( )	c. Hydrogen
D. Nitre	( )	d. Carbon dioxide
	( )	e. Nascent hydrogen

Directions for questions from 30 to 49

Answer the following questions.

30. Name the different layers of atmosphere.

31. What is atmosphere?

32. Name any two inert gases present in atmosphere.

33. Mention the percentage by volume of inactive and active components of atmosphere.

34. Name the device used to measure the atmospheric pressure.

35. How are the delicate colours of auroras caused?

36. Name the gas which absorbs ultraviolet rays.

37. Name the principal products of combustion of glucose.

38. Mention the significance of water vapour in atmosphere.

39. In which way is greenhouse effect useful to us?

40. Name the gases which are responsible for acid rain.

41. What happens when potassium nitrate is subjected to heating?

42. Define a catalyst.

43. Given the balanced chemical equation for the laboratory preparation of oxygen gas.

44. What is the action of litmus on oxygen?

45. What happens when a paper soaked in acidified potassium dichromate is exposed to SO_2 gas?

46. Mention the factors which influence rusting of iron.

47. What is alloying?

48. Name two metal carbonates from which carbon dioxide cannot be prepared by heating.

49. What is carbogen?

Short Answer Type Questions

50. Given any two chemical properties of CO_2.

51. What is greenhouse effect?

52. Mention various methods of minimising air pollution.

53. Explain the role of oxygen in the following:

(a) respiration (b) cutting and welding
(c) explosion

54. List out the important uses of CO_2.

55. Mention the ways of minimising global warming.

56. Mention the advantage of existence of atmosphere.

57. Explain ozone depletion.

58. Mention the effects of acid rain.

59. Mention various constituents of air and write their percentage by weight.

60. Mention any four physical properties of CO_2.

61. Write a short note on nascent hydrogen.

62. Explain hydrogenation of oils.

63. Mention any four important uses of hydrogen.

64. Explain the process of nitrogen fixation.

65. What is the composition of water gas and how is it formed?

66. Mention the action of oxygen with the elements sulphur, phosphorus, sodium and magnesium.

67. Mention the methods for preventing acid rain.

68. What are the human activities that contribute to air pollution?

Essay Type Questions

69. Describe the laboratory preparation of oxygen gas from hydrogen peroxide with a neat diagram.

70. Mention the general methods of preparation of CO_2 along with the chemical reactions.

71. Write a short note on rusting. What are the causes and effects of it? Explain different methods adopted to prevent rusting.

72. Explain the significance of nitrogen, oxygen and water vapour present in the atmosphere.

73. Describe the laboratory method of preparation of hydrogen gas.

74. Describe the chemical reaction of CO_2 with

(a) lime water
(b) metals like sodium, potassium and calcium
(c) non-metals like carbon

75. Write a short note on the following layers of atmosphere.

(a) troposphere (b) stratosphere

76. (a) What is air pollution?
(b) Write a short note on
(i) ozone depletion and
(ii) hazards of carbon monoxide

77. Fill in the blanks given in the following table.

Impurities	Substances used for removal
(a) AsH_3, PH_3	________
(b) ________	Lead nitrate solution
(c) ________	Caustic potash
(d) H_2O (water vapour)	________

78. Name the pollutants responsible for the following effects.

(a) reduction in intake of oxygen
(b) cause of acid rain
(c) greenhouse effect
(d) formation of smog
(e) depletion of ozone layer
(f) problems of digestion
(g) contamination of ground water
(h) respiratory disorders

For Hints and Explanations, please visit:
www.pearsoned.co.in/IITFoundationSeries

CONCEPT APPLICATION

Level 1

Directions for questions from 1 to 10

State whether the following statements are true or false.

1. Aeroplanes usually fly in the zone of mesosphere.
2. Meteorites burn up in thermosphere.
3. The percentage of noble gases present in air is more than that of CO_2.
4. Oxygen is neutral towards litmus.
5. Gases like nitrogen dioxide, sulphur dioxide present in atmosphere contribute to acid rain.
6. CO_2 can be used to preserve food grains from destruction by insects.
7. CO_2 can be collected by the downward displacement of air.
8. The electric spark produced during lightening helps nitrogen to combine with oxygen to form nitric oxide.
9. Carbogen is a mixture of 94 per cent oxygen, 4 per cent carbon dioxide and 2 per cent nitrogen dioxide.
10. Dust particles along with water vapour present in the atmosphere help in the formation of clouds.

Directions for questions from 11 to 20

Fill in the blanks.

11. ______ is present in sun's atmosphere up to 1.1 per cent.
12. ______ is used as catalyst in the hydrogenation of oils.
13. Composition of water gas is ________.
14. _______ is used as a rocket fuel.
15. Oxygen is _______ towards litmus.
16. Chemical formula of white lead is ________.
17. The atmospheric pressure is maximum at _____ and is found to be equal to _______.
18. In _____ layer of atmosphere TV waves do not get reflected. Hence, artificial satellites are used for television transmission.
19. Nascent hydrogen is a strong _______ agent. (reducing/oxidising)
20. ______is used for burning fuel in rockets and space ships.

Directions for questions from 21 to 37

For each of the questions, four choices have been provided. Select the correct alternative.

21. Nitrogen dilutes the activity of _____ in atmosphere.
    (a) Ar (b) CO_2
    (c) O_2 (d) He
22. Which among the following gases do not contribute to greenhouse effect?
    (a) methane (b) carbon dioxide
    (c) water vapour (d) nitrogen dioxide
23. _____ burns in oxygen with a brilliant golden yellow flame.
    (a) Na (b) Ca
    (c) Mg (d) K
24. The increase in the proportion of ______ in air leads to global warming.
    (a) CO (b) CO_2
    (c) SO_2 (d) N_2
25. The reaction of which of the following metal with steam is reversible?
    (a) calcium (b) aluminium
    (c) iron (d) lead
26. ______ is used in advertisement glow tubes.
    (a) neon (b) helium
    (c) argon (d) nitrogen
27. SI unit of atmospheric pressure is
    (a) Newton (b) Pascal
    (c) Joule (d) kg/m
28. Which among the following gas was called acid formerly?
    (a) nitrogen (b) hydrogen
    (c) carbon dioxide (d) oxygen
29. Ozone present in ______ plays a vital role in protecting the life on earth surface from harmful UV rays.
    (a) troposphere (b) stratosphere
    (c) mesosphere (d) ionosphere

30. Which among the following metal is used in alloying of iron to prevent corrosion?
(a) zinc (b) tin
(c) aluminium (d) chromium

31. Fountain ink pen leaks at higher altitudes. It is due to
(a) low atmospheric pressure
(b) high temperature
(c) high atmospheric pressure
(d) both (1) and (2)

32. The holes in the ozonosphere is due to the attack of a pollutant that is released from
(a) automobiles (b) air conditioners
(c) industries (d) volcanic eruptions

33. Which of the following phenomenon is an effect of acid rain?
(a) an increase in frequency of floods
(b) unseasonal rainfalls
(c) harmful effects by UV rays
(d) death of aquatic animals

34. Identify the substance which does not continue to burn in an atmosphere of N_2.
(a) Na (b) Mg
(c) K (d) W

35. A gas is collected by the upward displacement of air but cannot be collected either downward or upward displacement of water. Identify the probable characteristics of the gas.
(a) It is heavier than air and insoluble in water.
(b) Its vapour density is equal to air and is highly soluble in water.
(c) It is heavier than air and is highly soluble in water.
(d) It is lighter than air and water.

36. Arrange the layers of the atmosphere as per the following terms related in a sequence, respectively. Inversion layer, convection currents, auroras, absorbs harmful ionising radiations and burning up of meteorites
(1) troposphere (2) stratosphere
(3) exosphere (4) mesosphere
(5) ionosphere
(a) 45312 (b) 12354
(c) 21534 (d) 12435

37. Arrange the following given reactions in a sequence for the process of nitrogen fixation.
(1) formation of metal nitrates
(2) formation of nitrogen dioxide
(3) formation of nitric oxide
(4) formation of nitric acid
(a) 3214 (b) 3241
(c) 2143 (d) 4213

Directions for questions from 38 to 40

Match the entries given in Column A with the appropriate ones in Column B. More than one suitable content of Column B can also be considered correct for an entry in Column A and some content given in Column B may not be the appropriate one for any entry in Column A.

38.

Column A		Column B
A. Oxides of nitrogen	( )	a. Clogging drainage
B. Chloroflouro carbons	( )	b. Causes acid rain
C. Fly ash	( )	c. Formation of smog
D. Suspended particulate matter	( )	d. Depletion of ozone layer
	( )	e. Global warming

39.

Column A		Column B
A. $SO_2 + H_2O \rightarrow$	( )	a. H_2SO_4
B. $NO_2 + H_2O \rightarrow$	( )	b. H_2SO_3
C. $P_2O_5 + H_2O \rightarrow$	( )	c. H_3PO_4
D. $H_2O + CO_2$	( )	d. $H3PO_3$
	( )	e. H_2CO_3
	( )	f. $CO + H_2$ (water gas)
	( )	g. HNO_3
	( )	h. HNO_2

40.

Column A		Column B
A. Oxygen	( )	a. Preservation of food items
B. Nitrogen	( )	b. Good reducing agent
C. Carbon dioxide	( )	c. Kipps apparatus
D. Nascent hydrogen	( )	d. Dilutes the activity of oxygen
	( )	e. Major constituent of carbogen
	( )	f. Making of white lead

Level 2

Directions for questions from 1 to 20
For each of the questions, four choices have been provided.

Select the correct alternative

1. Meteorites burn up in the mesosphere but not in the thermosphere. It is due to
   (a) relatively thicker atmosphere in the mesosphere.
   (b) relatively thicker atmosphere in the thermosphere.
   (c) very low temperatures in the mesosphere.
   (d) very low temperatures in the thermosphere.

2. Aeroplanes usually fly in the stratosphere because
   (a) of the presence of ozone layer.
   (b) of the atmosphere being turbulent.
   (c) of the decrease in temperature with altitude.
   (d) of the absence of convection current.

3. Two non-metallic oxides A and B are produced by burning of fossil fuels. A pollutes both air and water. B does not cause water pollution and its presence in water is essential. It causes air pollution. Identify A and B.
   (a) NO_2, SO_2 (b) SO_2, CO
   (c) SO_2, CO_2 (d) CO_2, NO_2

4. Though zinc coating is cheaper than tin coating on iron, only tin coating is used for iron containers for storing food materials. Identify the correct reason.
   (a) Zinc is less reactive than tin.
   (b) Zinc acts as a preservative.
   (c) Tin does not react with the components of food material.
   (d) Tin lowers the ambient temperature, and thus, prevents spoilage of food.

5. An element X, in which the difference in electrons present in valence shell and penultimate shell is equal to the number of electrons present in the first shell, undergoes combustion to produce Y. Y turns lime water milky due to the formation of Z. Identify X, Y and Z.
   (a) C, CO_2, $CaCO_3$ (b) S, SO_2, $CaSO_3$
   (c) S, SO_2, $Ca(HSO_3)_2$ (d) C, CO_2, $Ca(HCO_3)_2$

6. A metal carbonate X when treated with dil HCl gives Y. 'Y' when used in mines area reduces suspended particulate matter in atmosphere. Identify Y.
   (a) NaCl (b) H_2CO_3
   (c) $CaCl_2$ (d) CO_2

7. The sculptures and monuments get damaged due to acid rains. Usage of scrubbers, containing lime, in industries decreases the percentage content of the major air pollutants that contribute to acid rain. The characteristic reaction involved in the scrubber is
   (a) neutralisation
   (b) oxidation
   (c) reduction
   (d) combination followed by decomposition

8. Air is a homogeneous mixture of gases which also contains argon and nitrogen which are used for filling electric bulbs. But air is not filled in electric bulbs because
   (a) the active component of air oxidises the filament.
   (b) the boiling point of air is greater than that of argon or nitrogen.
   (c) air has impurities like dust particles.
   (d) air exerts more pressure.

9. Both the formation of mist and clouds involve the condensation of water vapour. In which aspect do they differ from each other?
   (a) The process of cloud formation takes place in summer, whereas the process of mist formation takes place in winter.
   (b) The cloud formation takes place in lower layers while mist formation takes place in the upper layers of the troposphere.
   (c) Clouds are suspended in the atmosphere while mist settles down on leaves, walls, etc., on earth's surface.
   (d) The water droplets of a cloud have a much lower temperature than that of water droplets of mist.

10. The CO that is added up to the atmosphere due to the incomplete combustion of fuels can be decreased by the
    (a) usage of scrubber.
    (b) usage of electric precipitator.
    (c) usage of catalytic convertor.
    (d) increasing greenery.

11. Y is obtained by the decomposition of a substance X. Oxygen present in X is univalent. Y reacts with a neutral oxide and evolves Z which is composed of reddish brown fumes which turns moist blue litmus to red. Identify X, Y and Z.
    (a) H_2O_2, H_2 and CO_2 (b) H_2O_2, O_2 and NO_2
    (c) $KClO_3$, Cl_2 and O_2 (d) HgO, O_2 and NO

12. Assume that temperature is recorded in the thermosphere with the help of a thermometer. You will be amazed to find that the temperature recorded in it will be extremely low. This is because
    (a) of the complete absorption of high energetic radiations by molecules.
    (b) of the absence of atmosphere.
    (c) of the presence of very few molecules or ions.
    (d) it is at very high altitudes.

13. Cloud formation almost does not take place in the stratosphere because
    (a) the high temperature of this layer does not allow the water vapour to reach this layer.
    (b) of the absence of convection current.
    (c) of the absence of dust particles and convection current.
    (d) all of the above

14. Which of the following pairs is a major contributor to global warming?
    (a) carbon dioxide, nitrogen oxide
    (b) methane, water vapour
    (c) carbon dioxide, ozone
    (d) carbon dioxide, water vapour

15. Generally, a metal gets corroded due to oxidation. Name a metal which becomes chemically resistant due to oxidation.
    (a) zinc (b) copper
    (c) aluminium (d) iron

16. An element X burns with a brilliant blue flame in the presence of oxygen to form Y. Y turns acidified potassium dichromate paper to green. Identify X and Y.
    (a) nitrogen, nitrogen dioxide
    (b) sulphur, sulphur dioxide
    (c) phosphorus, phosphorus pentoxide
    (d) sodium, sodium oxide

17. Which of the following processes does not contribute to the increase of suspended particulate matter in the atmosphere?
    (a) emissions from industries
    (b) burning of coal
    (c) burning of wood
    (d) usage of refrigerators and air conditioners

18. Incomplete combustion of fuels does not contribute to
    (a) photochemical smog (b) acid rain
    (c) global warming (d) both (a) and (b)

19. Leguminous plants do not require the addition of nitrogenous fertilisers because
    (a) they can fix up atmospheric nitrogen.
    (b) of the presence of rhizobium bacteria in their root nodules.
    (c) they possess nitrogen in roots, and hence, do not require external intake.
    (d) both (1) and (2)

20. A gas: X when passed through blue litmus solution turns it to red. Another gas Y when passed through either blue litmus or red litmus does not change the colour of the litmus. Identify the true statement regarding X and Y.
    (a) X is basic while Y is neutral.
    (b) Y is neutral while X is acidic.
    (c) Y is basic while X is acidic.
    (d) Y is acidic while X is neutral.

21. While a teacher was teaching her class about air and oxygen she emphasised the importance of nitrogen in the atmosphere. A student named Vivek asked the teacher, What would happen if the percentages of nitrogen and oxygen were interchanged in the atmosphere? The teacher praised him for the question and explained it. What was the teacher's explanation?

22. Vish went to an electrical shop along with her father to purchase a bulb and some electrical equipment. After returning home, Vish looking at the bulb keenly asked her father. Is it filled with air or vacuum? Her father told her that it was filled with nitrogen or argon. She asked him why the bulb could not be filled with air which is readily available, as nitrogen is obviously a part of air? What was her father's correct explanation to her question?

23. Vidya went to her grandfather Rajaram's house to spend her summer vacation. Rajaram has coconut fields near his house. In the evening she went to the coconut field along with her grandfather. There the workers offered her coconut water. Her grandfather told the workers to provide a straw to facilitate drinking the coconut water comfortably. Can you explain the principle behind using straw for the above purpose?

24. Shakti along with his parents was going to the railway station by an auto to catch a train. Shakti's father was recollecting his childhood days and said that in those days, autos were run with petrol only. His mother added that diesel was used to run autos in later days. Shakti said that nowadays autos are preferably run with CNG. Shakti questioned his father regarding the advantage of CNG over petrol and diesel. What was his father's explanation?

25. Arjun's family stays in an area which is in the heart of the city for the sake of convenience. During the summer vacation, Arjun's family stays in their farm house which has all amenities. As the farm house is located in the city outskirts, they could escape from the hot sun in summer. Why is the temperature in the peripheral area relatively lower than that in the heart of the city?

26. Calculate the volume of nitrogen, oxygen and carbon dioxide in 750 L of air.

27. Troposphere is extended up to 6–20 km altitude from the earth's surface but the amount of oxygen is less at higher altitudes, i.e., on mountains or hill tops. Justify.

28. X and Y are gases at room temperature. X is less reactive and is used to preserve food material, while Y is required for combustion. Metal M gets corroded when it combines with Y in the presence of Z. An atom of M possesses 14 electrons in the second last shell and N is the last shell. Identify X, Y, Z and M.

29. Both galvanising and tinning are adopted for preventing rusting of iron but only tinned iron containers are used to preserve food material. Justify.

30. Why are people advised not to sleep in a closed room where coal is burnt in chulhas?

31. The compounds X and Y combine to form a compound Z. 'Z' which is used for construction of monuments gets affected by acid rain. The compound X is used in white washing. Identify X, Y and Z and write the corresponding balanced chemical equations.

32. What is the role of water vapour in the formation of fog?

33. An inverted glass jar is placed over a burning candle. After a while, the candle is removed and the inverted glass jar is placed in a trough of water. What do you observe and justify your observations.

34. Acid formed as a result of lightening reaches the earth's surface through precipitation, thus contributing to the acid rain as well as to the growth of plants. Justify.

Level 3

1. A, B, C and D are four layers of atmosphere. A helps in long-distance radio communications. In layer B, most of weather phenomena take place. The pollutants released from air conditioners and aerosols deplete layer D which also gets depleted by nitrous oxide. Meteorites burn up in layer C. Identify the layers A, B, C and D and arrange them in an increasing order of height from the surface of earth.

2. The element 'X' has only three shells in which the difference in number of electrons between the last shell and the second last shell is equal to the

PRACTICE QUESTIONS

number of electrons in the first shell. This element 'X' burns in oxygen to produce compound Y, which turns acidified potassium dichromate paper to green. Y turns moist blue litmus paper to red due to the formation of compound Z. Identify X, Y and Z.

3. Ozone present in the troposphere is a major pollutant. However, ozone formation in aqua purifiers makes drinking water pure and safe. Explain.

4. Carbon dioxide is used for extinguishing fires caused by petroleum products but not fires caused by metals like Na or Mg. Explain.

5. Explain the principle behind the usage of electric precipitators in industries and catalytic converters in automobiles for controlling air pollution.

ASSESSMENT TEST

Test 1

Directions for questions from 1 to 15

Select the correct alternative from the given choices.

1. Identify the component which when present in the troposphere is a pollutant and when present in upper layers of atmosphere is a protector of life.
   (a) chloroflurocarbons
   (b) ozone
   (c) nitrous oxide
   (d) water vapour

2. Y is obtained by the decomposition of a substance X. Oxygen present in X is univalent. Y reacts with a neutral oxide and evolves Z which is composed of reddish brown fumes which turns moist blue litmus red. Identify X, Y and Z.
   (a) H_2O_2, H_2 and CO_2
   (b) H_2O_2, O_2 and NO_2
   (c) $KClO_3$, Cl_2 and O_2
   (d) HgO, O_2 and NO

3. Identify the four layers of atmosphere as per the temperature range mentioned and arrange them in a sequence.

   −60°C to −15°C, 17°C to −51°C, −15°C to −120°C and maximizing up to 2000°C
   (1) troposphere (2) stratosphere
   (3) mesosphere (4) exosphere
   (5) thermosphere
   (a) 2 1 3 5 (b) 1 2 3 5
   (c) 2 1 3 4 (d) 3 1 2 5

4. Match the following:

Column A		Column B
A. Coating with	( )	a. Bicycle handles, rims red lead oxide
B. Coating with enamel paint	( )	b. Bridges, electric poles
C. Electroplating	( )	c. Surgical instruments, cutlery
D. Alloying	( )	d. Bodies of cars, buses

   (a) A → b, B → d, C → a, D → c
   (b) A → b, B → c, C → d, D → a
   (c) A → a, B → d, C → b, D → c
   (d) A → d, B → c, C → a, D → b

5. Air is a homogeneous mixture of gases which also contains argon and nitrogen which are used for filling electric bulbs. But air is not filled in electric bulbs because
   (a) the active component of air oxidises the filament.
   (b) the boiling point of air is greater than that of argon or nitrogen.
   (c) air has impurities like dust particles.
   (d) air exerts more pressure.

6. A list of four different substances is given below.

   The methods of preparation of the substances are also given. Arrange the methods of preparation in the order of the substances given.
   (i) O_2 (ii) CO_2
   (iii) SO_2 (iv) P_2O_5

(1) burning of a non-metal produces dense white fumes.

(2) burning of a non-metal with sparkles and crackling sound.

(3) heating of one of the binary compound of hydrogen.

(4) burning of a non-metal with a brilliant blue flame.

(a) 3 2 4 1 (b) 2 3 4 1

(c) 1 4 3 2 (d) 3 4 1 2

7. Excess intake of O_2 by a person results in several ill-effects. Which of the following activities leads to excess intake of O_2?

(a) deep-sea diving

(b) mountaineering

(c) travelling in an aeroplane

(d) walking on the moon

8. **Assertion (A)**: Hydrogen gas is one of the components in both oxyhydrogen flame and oxyacetylene flame.

Reason (R): Hydrogen is inflammable.

(a) Both (A) and (R) are true and (R) is the correct reason of (A).

(b) Both (A) and (R) are true but (R) is not the correct reason of (A).

(c) (A) is true but (R) is false.

(d) (A) is false but (R) is true.

9. The sculptures and monuments get damaged due to acid rains. Usage of scrubbers, containing lime water, in industries decreases the percentage content of the major air pollutants that contribute to acid rain. The characteristic reaction involved in the scrubber is

(a) neutralisation

(b) oxidation

(c) reduction

(d) combination followed by decomposition

10. An element X, in which the difference in electrons present in valence shell (M shell) and penultimate shell is equal to the number of electrons present in the first shell, undergoes combustion to produce Y. Y turns lime water milky due to the formation of Z. Identify X, Y and Z.

(a) C, CO_2, $CaCO_3$

(b) S, SO_2, $CaSO_3$

(c) S, SO_2, $Ca(HSO_3)_2$

(d) C, CO_2, $Ca(HCO_3)_2$

11. Though zinc coating is cheaper than tin coating on iron, only tin coating is used for iron containers for storing food materials. Identify the correct reason.

(a) Zinc is less reactive than tin.

(b) Zinc acts as a preservative.

(c) Tin does not react with the components of food material.

(d) Tin lowers the ambient temperature, and thus, prevents spoilage of food.

12. Which of the following can be used to reduce suspended particulate matter in atmosphere in mine areas?

(a) NaCl

(b) CaO

(c) $CaCl_2$

(d) All the above

13. **Assertion (A)**: If excess of carbon dioxide is passed through lime water, it becomes transparent.

Reason (R): Calcium bicarbonate is soluble in water.

(a) Both (A) and (R) are true and (R) is the correct explanation for (A).

(b) Both (A) and (R) are true but (R) is not the correct explanation for (A).

(c) (A) is true but (R) is false.

(d) (A) is false but (R) is true.

14. O_2 and CO_2 prepared in the laboratory can be collected by ________.

(a) downward displacement of water and downward displacement of air, respectively.

(b) downward displacement of water and upward displacement of air, respectively.

(c) upward displacement of water in both the cases.

(d) downward displacement of water in both the cases.

15. Two non-metallic oxides A and B are produced by burning of fossil fuels. A pollutes both air and water. B does not cause water pollution and its presence in water is essential. It causes air pollution. Identify A and B.

(a) NO_2, SO_2 (b) SO_2, CO

(c) SO_2, CO_2 (d) CO_2, NO_2

Test 2

Directions for questions from 1 to 15

Select the correct alternative from the given choices.

1. The holes in the ozonosphere is due to the attack of a pollutant that is released from
   (a) automobiles (b) air conditioners
   (c) industries (d) volcanic eruptions

2. A gas X when passed through blue litmus solution turns it to red. Another gas Y when passed through either blue litmus or red litmus does not change the colour of the litmus. Identify the true statement regarding X and Y.
   (a) X is basic while Y is neutral
   (b) Y is neutral while X is acidic
   (c) Y is basic while X is acidic
   (d) Y is acidic while X is neutral

3. Identify the layers of atmosphere as per activities mentioned and arrange them in a sequence. Burning up of meteorites, flying of aeroplanes, reflection of radio waves, absorption of harmful ionising radiations and occurrence of most of the weather phenomena.
   (1) troposphere
   (2) stratosphere
   (3) mesosphere
   (4) ionosphere
   (5) magnetosphere
   (a) 2 3 4 5 1 (b) 3 2 4 5 1
   (c) 3 4 5 1 2 (d) 2 4 1 5 3

4. Match the following:

Column A		Column B
A. Tinning	( )	a. Iron trunks, sanitary
B. Enamelling	( )	b. Engine parts
C. Coating with oil, grease	( )	c. Wash basins, ovens
D. Galvanisation	( )	d. Articles used to preserve edible food materials

   (a) A → c, B → a, C → b, D → d
   (b) A → d, B → c, C → b, D → a
   (c) A → d, B → c, C → a, D → b
   (d) A → b, B → d, C → a, D → c

5. Identify the substance which continues to burn in an atmosphere of N_2.
   (a) Fe (b) Mg
   (c) Cu (d) W

6. Characteristic reactions for four different substances are given below. Arrange the tests in the order of the substances given below.
   (i) P_2O_5 (ii) CO_2
   (iii) O_2 (iv) SO_2
   (1) Passage through potassium dichromate turns green.
   (2) Dissolution in water gives a tribasic acid.
   (3) Non-supporter of combustion but allows active metals continue to burn.
   (4) Turns alkaline pyrogallol brown.
   (a) 2 4 3 1 (b) 2 3 4 1
   (c) 4 1 3 2 (d) 4 3 1 2

7. Fountain ink pen leaks at higher altitudes. It is due to
   (a) low atmospheric pressure
   (b) high temperatures
   (c) high atmospheric pressure
   (d) both (1) and (2)

8. **Assertion (A)**: Many fuel gases contain hydrogen as a major constituent.

   Reason (R): Liquid hydrogen is a very light fuel.
   (a) Both (A) and (R) are true and (R) is the correct explanation for (A).
   (b) Both (A) and (R) are true but (R) is not the correct explanation for (A).
   (c) (A) is correct but (R) is false.
   (d) (A) is false but (R) is true.

9. Which of the following phenomena is an effect of acid rain?
   (a) an increase in frequency of floods
   (b) unseasonal rainfalls
   (c) harmful effects by UV rays
   (d) death of aquatic animals

10. An element X burns with a brilliant blue flame in the presence of oxygen to form Y. Y turns acidified potassium dichromate paper green. Identify X and Y.

PRACTICE QUESTIONS

(a) nitrogen, nitrogen dioxide
(b) sulphur, sulphur dioxide
(c) phosphorus, phosphorus pentoxide
(d) sodium, sodium oxide

11. Generally, a metal gets corroded due to oxidation. Name a metal which becomes chemically resistant due to oxidation.
(a) zinc
(b) copper
(c) aluminium
(d) iron

12. Which of the following processes does not contribute to the increase of suspended particulate matter in the atmosphere?
(a) emissions from industries
(b) burning of coal
(c) burning of wood
(d) usage of refrigerators and air conditioners

13. **Assertion (A)**: When carbon dioxide gas is passed through lie water, lime water is turned milky.

Reason (R): Calcium carbonate is soluble in water.
(a) Both (A) and (R) are true and (R) is correct explanation for (A).
(b) Both (A) and (R) are true but (R) is not the correct explanation for (A).
(c) (A) is correct but (R) is false.
(d) (A) is false but (R) is true.

14. A gas is collected by the upward displacement of air but cannot be collected by downward displacement of water. Identify the probable characteristics of the gas.
(a) It is heavier than air and insoluble in water.
(b) Its vapour density is equal to air and it is highly soluble in water.
(c) It is heavier than air and it is highly soluble in water.
(d) It is lighter than air and water.

15. Which of the following pairs is a major contributor to global warming?
(a) carbon dioxide and oxides of nitrogen
(b) methane and water vapour
(c) carbon dioxide and ozone
(d) carbon dioxide and water vapour

TEST YOUR CONCEPTS

Fill in the blanks

1. ozonosphere
2. thermosphere
3. Chlorofluorocarbons
4. catalyst
5. MnO_2
6. Carbon dioxide
7. dry ice
8. Helium
9. $Fe_2O_3.XH_2O$
10. greenhouse effect

Multiple choice questions

11. b
12. b
13. c
14. c
15. a
16. c
17. b
18. b
19. b
20. b
21. a
22. b
23. b
24. d
25. d
26. a

Match the following

27. A → d, e
B → a
C → b
D → c

28. A → b
B → d
C → a, e
D → a, e

29. A → b
B → d
C → c
D → b

CONCEPT APPLICATION

Level 1

True or false

1. False
2. False
3. True
4. True
5. True
6. True
7. False
8. True
9. False
10. True

Fill in the blanks

11. Hydrogen
12. Nickel
13. $CO + H_2$
14. Liquid hydrogen
15. neutral
16. $Pb(OH)_2.PbCO_3$
17. sea level, 760 mm of Hg
18. Ionosphere
19. reducing
20. liquid oxygen

Multiple choice questions

21. c
22. d
23. a
24. b
25. c
26. a
27. b
28. d
29. b
30. d
31. a
32. b
33. d
34. d
35. c
36. c
37. b

Match the following

38. A → b, c, e
B → d, e
C → a
D → c

39. A → b
B → g, h
C → c
D → e

40. A → e
B → d, a
C → f, a
D → b, c

Level 2

Multiple choice questions

1. a	2. d	3. c	4. c	5. b
6. c	7. a	8. a	9. b	10. c
11. b	12. c	13. d	14. d	15. c
16. b	17. d	18. c	19. d	20. b

ASSESSMENT TEST

Test 1

1. b	2. b	3. a	4. a	5. a
6. a	7. a	8. d	9. a	10. b
11. c	12. c	13. a	14. b	15. c

Test 2

1. b	2. b	3. b	4. b	5. b
6. b	7. a	8. b	9. d	10. b
11. c	12. d	13. c	14. c	15. d

CONCEPT APPLICATION

Level 1

True or false

1. Aeroplanes usually fly in the zone of stratosphere.
2. Meteorites burn up in mesosphere.
3. The percentage of noble gases is 0.94 per cent and that of CO_2 is 0.03 per cent in atmosphere.
4. Oxygen is neutral towards litmus as it is neither acidic nor basic.
5. Gases like nitrogen dioxide, sulphur dioxide in atmosphere cause acid rain.
6. When food grains are kept in the atmosphere of carbon dioxide, it prevents the grain from destruction by insects.
7. CO_2 can be collected by upward displacement of air as CO_2 is denser than air.
8. It provides enough temperature for nitrogen and oxygen to combine to form nitric oxide.
9. Carbogen is a mixture of 95 per cent oxygen and 5 per cent carbon dioxide.
10. Dust particles help in the formation of cloud.

Fill in the blanks

11. Hydrogen is present in sun's atmosphere up to 1.1 per cent.
12. Nickel is used as a catalyst in the hydrogenation of oils.
13. Composition of water gas is $CO + H_2$.
14. Liquid hydrogen is used as a rocket fuel.
15. Oxygen is neutral towards litmus.
16. Chemical formula of white lead is $Pb(OH)_2 \cdot PbCO_3$.
17. The atmospheric pressure is maximum at sea level and is found to be equal to 760 mm of Hg.
18. In the ionosphere, a layer of atmosphere, TV waves do not get reflected. Hence, artificial satellites are used for transmission of TV waves.
19. Nascent hydrogen is a strong reducing agent.
20. Liquid oxygen is used for burning fuels in rockets and space ships.

Multiple choice questions

21. Nitrogen dilutes the activity of oxygen in the atmosphere.
22. Nitrogen dioxide present in the atmosphere does not contribute to acid rain.
23. Sodium burns in oxygen with a brilliant golden yellow flame.
24. The increase in the proportion of CO_2 in air leads to global warming.
25. The reaction of iron with steam is a reversible reaction.
26. Neon is used in advertisement glow tubes.
27. SI unit of atmospheric pressure is Pascal.
28. Oxygen gas is called acid former.
29. Ozone present in the stratosphere plays a vital role in protecting the life on earth's surface from harmful UV rays.
30. Chromium metal is used in the alloying of iron to prevent corrosion.
31. As we move to higher altitudes, fountain ink pen leaks out because of lower atmospheric pressure. The pressure built up in the ink pen is more when compared to the outside pressure. Hence, it leaks out.
32. Holes are created in the ozone layer due to the attack of CFCs which are released from refrigerators and air conditioners.

33. Due to acid rain, sculptures and monuments get damaged and also water bodies like rivers, oceans get acidified. This acidified water affects the aquatic life which leads to death and decay of aquatic animals.

34. Since sodium, magnesium and potassium are active metals, they catch fire when they react with oxygen and continue to burn in an atmosphere of N_2, while tungsten (W) which burns in presence of oxygen, does not continue to burn in an inert atmosphere of N_2.

35. 'X' is highly soluble in water. Hence, it cannot be collected by water. It is collected by the upward displacement of air as it is heavier than air.

36. (i) Stratosphere (ii) Troposphere
(iii) Ionosphere (iv) Exosphere
(v) Mesosphere

37. (i) Formation of nitric oxide
(ii) Formation of nitrogen dioxide
(iii) Formation of nitric acid
(iv) Formation of metal nitrates

Match the following

38. A → b, c, e Oxides of nitrogen contribute to acid rain, global warming and formation of smog.
B → d, e CFC contributes to the depletion of the ozone layer and global warming
C → a fly ash causes clogging of drainage.
D → c Suspended particulate matter involved in the formation of smog.

39. A → b $SO_2 + H_2O \rightarrow H_2SO_3$
B → g, h $2NO_2 + H_2O \rightarrow HNO_2 + HNO_3$
C → c $P_2O_5 + 3H_2O \rightarrow 2H_3PO_4$
D → e $CO_2 + H_2O \rightarrow H_2CO_3$

40. A → e The major constituent of carbogen is oxygen.
B → d, a Nitrogen dilutes the activity of oxygen. Nitrogen has an inert nature, hence, it is used to preserve food items.
C → f, a CO_2 is used in the making of white lead. CO_2 is also used in the preservation of food stuffs.
D → b, c Nascent hydrogen acts as a good reducing agent. Nascent hydrogen can be replenished by using Kipps apparatus.

Level 2

1. Meteorites burn up in the mesosphere due to the presence of relatively thick atmosphere. Meteorites collide with molecules present in this layer, and thus emit, light known as shooting stars. While in the thermosphere, the thin atmosphere is not enough to produce friction for the meteorites to burn.

2. Aeroplanes fly in the stratosphere due to the absence of convection current. This is because the temperature gradually increases with increase in altitude due to the absorption of highly energetic UV radiations. Since there is no turbulence in the atmosphere of the stratosphere, aeroplanes usually fly in the stratosphere.

3. 'A' can be SO_2 or NO_2 because they pollute air and cause acid rain to pollute water. 'B' can be CO_2 since it does not cause acid rain. Its presence in water is essential for aquatic plants. It is considered as an air pollutant as it results in global warming.

4. Tin does not react with the components of food material while Zn reacts and leads to spoilage of food. Hence, tin-coated containers are preferred over zinc-coated ones as tin is less reactive than Zn.

5. Electronic configuration X is of 2, 8, 6.

Hence, X is sulphur.

$$\underset{(X)}{S} + O_2 \rightarrow \underset{(Y)}{SO_2}$$

$$\underset{(Y)}{SO_2} + Ca(OH)_2 \rightarrow \underset{(Z)}{CaSO_3} \rightarrow + H_2O$$

(milky white)

6. Anhydrous calcium chloride is deliquescent in nature which absorbs moisture and forms a solution. Thus, $CaCl_2$ solution formed does not allow the dust particles and other suspended particulate matter to rise.

$$\underset{(X)}{CaCO_3} + HCl \rightarrow \underset{(Y)}{CaCl_2} + CO_2 + H_2O$$

7. Marble is made of $CaCO_3$ which gets dissolved when it reacts with acids like H_2SO_4 or HNO_3 of acid rains. The major pollutants which contribute to acid rain NO_2 and SO_2 are released into the atmosphere. The NO_2 and SO_2 percentage content in the atmosphere can be decreased by using equipment like scrubbers. In a scrubber, powdered $CaCO_3$ is taken which undergoes decomposition to give CaO. This quick lime combines with SO_2 to form $CaSO_3$ which is removed as slurry.

$$\underset{\text{Base}}{CaO} + \underset{\text{acid}}{SO_2} \rightarrow CaSO_3\downarrow$$

Hence, it is a neutralisation reaction.

8. Since air consists of an active component that is oxygen, it oxidises tungsten filament or any other metal used as a filament. Hence, electric bulbs are filled with inert gas like argon or chemically inert nitrogen.

9. Formation of mist takes place in lower layers while cloud formation takes place in the upper layers of the troposphere.

The process of cloud formation takes place throughout the year while the formation of mist takes place only at certain times at particular temperatures.

10. Catalytic convertors consist of ceramic honey comb like structures with platinum catalyst spread out. The increased surface area of the catalyst provides many reactive sites where the less oxidised products like CO and NO are converted to CO_2 and N_2 which are released into the atmosphere.

11. $$\underset{(X)}{H_2O_2} \rightarrow H_2O_2 + O_2$$

$$\underset{(Y)}{NO} + O_2 \rightarrow \underset{\substack{(Z)\\ \text{(reddish brown)}}}{NO_2}$$

O_2 (Y) is neutral towards litmus while NO_2 is acidic, and hence, turns moist blue to red.

12. Due to a very thin atmosphere in the thermosphere, only a few molecules present here absorb high energetic radiations, and hence, the temperature of molecules or ions formed is great but the ambient temperature is very low.

13. Due to the absorption of highly energetic radiations, i.e., UV radiations by the stratosphere, the temperature is high when compared to the troposphere. The high temperature of the stratosphere does not allow the water vapour to reach the stratosphere due to the absence of convection current. Hence, cloud formation does not take place in the stratosphere. Cloud formation requires convection currents and also dust particles for condensation of water droplets.

14. CO_2 and water vapour are greenhouse gases present in the atmosphere. Their percentage increase in the atmosphere increases the temperature of the earth and contributes to global warming. These are the major contributors to global warming.

15. Aluminium, when exposed to the atmosphere, reacts with air, i.e., O_2, and forms Al_2O_3 which is chemically inert. This layer of Al_2O_3 (alumina) makes aluminium passive.

16. Sulphur is the non-metal which burns with a blue-coloured flame. SO_2 turns orange-coloured $K_2Cr_2O_7$ to green due to the formation of $Cr_2(SO_4)_3$.

17. The usage of refrigerators and air conditioners releases chlorofluoro carbons which does not contribute to SPM.

18. Incomplete combustion of fuels releases CO, NO, etc., that leads to photochemical smog and acid rain but not global warming.

19. Leguminous plants consist of nodules in their roots with rhizobium bacteria. These bacteria convert atmospheric nitrogen into its corresponding metal nitrates. These nitrates can be readily assimilated by plants. Hence, they do not require additional nitrogenous fertilisers.

20. 'X' is an acidic gas since it changes blue litmus to red. 'Y' is neutral since it does not respond to either red or blue litmus.

21. Increased levels of oxygen increase the rate of combustion. Fuel would burn at a faster rate and become dangerous and it will be difficult to extinguish fires. It also would increase the rate of corrosion of metals and increase the rate of respiration which would affect the human system.

22. Air cannot be filled in electric bulbs as air contains oxygen which oxidises the filament. Hence, inert gas like Ar or less reactive gases like nitrogen are used as it does not react with the metal filament.

23. When we suck one end of a straw dipped into the coconut water, some air from inside the straw is drawn out. A partial vacuum is created within the straw and as a result coconut water is pushed up by the atmospheric pressure to fill up the vacuum and reaches the persons mouth. Hence, we can drink the coconut water.

24. CNG does not produce any harmful gas, and thus, it reduces the maintenance cost and it is a clean fuel. While petrol and diesel contain N and S which upon combustion produce NO_2 and SO_2, respectively. These lead to acid rain. Hence, CNG can be a substitute for petrol and diesel.

25. In the interior of cities, the density of population is more, as a result a large volume of traffic exists which leads to an increase in the gaseous contents such as CO_2, methane and nitrous oxides which contribute to global warming. Hence, the temperature in the interior of the cities is higher than in the outskirts of the cities.

26. In 100 L of air, volume of nitrogen = 78.10 L

$$\text{Volume of nitrogen in 750 L} = \frac{78.1 \times 750}{100}$$
$$= 585.75 \text{ L}$$

Volume of oxygen in 100 L of air = 21 L

$$\text{Volume of oxygen in 750 L} = \frac{21 \times 750}{100}$$
$$= 157.5 \text{ L}$$

Volume of CO_2 in 100 L of air = 0.03 L

$$\text{Volume of } CO_2 \text{ in 750 L} = \frac{0.03 \times 750}{100}$$
$$= 0.225 \text{ L}$$
$$= 225 \text{ mL}$$

27. The vapour density of oxygen is 16 and that of the air is 14.4. Hence, due to higher vapour density of oxygen its amount in air decrease on higher altitudes such as mountains, i.e., more denser molecules experience more gravitational pull.

28. 14 electrons in the M shell and 2 electrons in the N shell correspond to the metal 'Fe.' X is nitrogen which is filled in air tight food packets to prevent oxidation and Y is oxygen which is required for combustion. In presence of oxygen and water vapour (Z) metal iron (M) gets rusted to form a compound, i.e., $Fe_2O_3 \cdot XH_2O$. X = nitrogen Y = oxygen, Z = water vapour and M = Fe.

29. In the process of galvanising, zinc is coated on the iron. Zinc can react with acids present in food material and as a result food items cannot be consumed. However, tin cannot react with acids present in food items. Hence, tinned iron sheets are extensively used for making boxes to store edible substances.

30. It is dangerous to sleep in a closed room with coke burning in improperly designed chulhas because coal undergoes incomplete combustion and produces carbon monoxide which is a harmful gas and leads to respiratory problems.

31. Since X is used in white washing, X is $Ca(OH)_2$, Y is CO_2 and Z is $CaCO_3$ as it is used for construction of monuments.

$$\underset{X}{Ca(OH)_2} + \underset{Y}{CO_2} \rightarrow \underset{Z}{CaCO_3} + H_2O$$

$$CaCO_3 + H_2O + CO_2 \rightarrow Ca(HCO_3)_2.$$

32. The presence of water vapour in atmosphere influences the climatic conditions. Water vapour present in the atmosphere gets condensed on dust particles during night time (winter nights) because of low temperature. Depending upon the temperature, the water vapour produces snow, fog, mist, hails and other phenomena. When water vapour condenses on dust particles and remains suspended in atmosphere, it is called fog.

33. Burning of candle produces CO_2 and water. When the cylinder containing CO_2 is inverted into water, CO_2 dissolves in water and as a result the water level slightly rises in the glass jar. Thus, it indicates that CO_2 is soluble in water.

34. During acid rain, the nitric acid formed as a result of lightening process in very dilute form is washed down to earth and reacts with metal carbonates to form nitrates.

$CaCO_3 + 2HNO_3 \rightarrow Ca(NO_3)_2 + CO_2 + H_2O$. Nitrates so formed act as excellent nutrients for the growth of plants.

Level 3

1. A = ionosphere or thermosphere, B = troposphere, C = mesosphere, D = stratosphere. Increasing order of five layers as per altitude from the surface of earth are B < D < C < A.

2. The element X is sulphur which burns in oxygen to produce compound Y which (SO_2) turns acidified potassium dichromate paper to green. When moist blue litmus paper is introduced to SO_2, it turns red due to formation of compound Z which is H_2SO_3.

$$S + O_2 \rightarrow SO_2;\ SO_2 + H_2O \rightarrow H_2SO_3$$

$$SO_2 + K_2Cr_2O_7\ H_2SO_4 \rightarrow K_2SO_4 + \underset{\text{green}}{Cr_2(SO_4)} + H_2O$$

3. Aqua purifiers consist of energy source which radiates UV rays. Oxygen is converted to ozone in presence of ultraviolet rays. The ozone decomposes into oxygen and nascent oxygen in presence of UV rays. The nascent oxygen has ability to kill the harmful microorganisms present in water, and hence, water gets purified.

$$3O_2 \rightarrow 2O_3$$

$$O_3 \rightarrow O_2 + [O]$$

4. Active metals like Na and Mg react with carbon dioxide and form corresponding oxide and carbon monoxide in trace amounts which is a harmful gas and also a supporter of combustion. Hence, fires caused by metals like Na and Mg cannot be extinguished by CO_2.

5. Electric precipitator removes the suspended particulate matter which are charged particles coming from thermal power plants, cement plants, steel plants so that smoke and ash entering the atmosphere is minimised. Catalytic converter contains catalysts such as platinum on ceramic comb which convert harmful gases to less harmful gases, such as CO to CO_2 and oxides of nitrogen to nitrogen.

ASSESSMENT TEST

Test 1

1. Ozone when present in the troposphere is a major pollutant that leads to global warming, while ozone present in the stratosphere acts as a shield from UV rays reaching the earth's surface.

2. $\underset{(X)}{H_2O_2} \rightarrow \underset{(Y)}{H_2O_2} + O_2$

$$NO + \underset{(Y)}{O_2} \rightarrow \underset{\substack{(Z)\\ \text{(reddish brown)}}}{NO_2}$$

O_2 (Y) is neutral towards litmus while NO_2 is acidic, and hence, turns moist blue to red.

3. Stratosphere has a temperature range of −60°C to −15°C, troposphere has a temperature range from 17°C to −51°C, mesosphere has a temperature range from −15°C to −120°C and thermosphere has a maximum temperature up to 2000°C. bace.

4. A → b Bridges and electric poles are coated with red lead oxide to prevent corrosion.

B → d Bodies of cars and buses are coated with enamel paint to prevent corrosion.

C → a Bicycle handles, rims are electroplated to avoid rusting.

D → c Surgical instruments and cutlery are made of alloys in order to prevent corrosion.

5. Since air consists of an active component, i.e., oxygen, it oxidises tungsten filament or any other metal used as a filament. Hence, electric bulbs are filled with an inert gas like argon or chemically inert nitrogen.

6. (i) Oxygen is prepared by heating of hydrogen peroxide.

(ii) Carbon burns with sparkles and cracking sound to give CO_2.

(iii) Sulphur burns with a brilliant blue flame to give SO_2.

(iv) P_2O_5 appears as dense white fumes when phosphorus is burnt with oxygen.

7. Due to increased pressure in the deep sea, more amount of oxygen gets dissolved in blood. This excessive intake of oxygen by a person affects and leads to several problems like nausea, dizziness, etc. Hence, a cylinder in which oxygen is diluted with an inert gas like He has to be carried.

8. In oxyhydrogen flame, hydrogen gas is involved due to its inflammable nature.

9. Marble is made of $CaCO_3$ which gets dissolved when it reacts with acids like H_2SO_4 or HNO_3 of acid rains. The major pollutants which contribute to acid rain (NO_2 and SO_2) are released into the atmosphere. The NO_2 and SO_2 percentage content in the atmosphere can be decreased by using equipment like scrubbers. In a scrubber, lime water is taken which combines with SO_2 to form $CaSO_3$ which is removed as slurry. Hence, it is a neutralisation reaction.

$$\underset{\text{base}}{Ca(OH)_2} + \underset{\text{acid gas}}{SO_2} \rightarrow CaSO_3 + H_2O$$

10. Electronic configuration of X is 2, 8, 6

Hence, X is sulphur.

$$\underset{(X)}{S} + O_2 \rightarrow \underset{(Y)}{SO_2}$$

$$\underset{Y}{SO_2} + Ca(OH)_2 \rightarrow \underset{\text{(milky white)}}{CaSO_3\downarrow} + \underset{Z}{H_2O}$$

11. Tin does not react with the contents of food material while Zn reacts and leads to spoilage of food. Hence, tin-coated containers are preferred over zinc-coated ones as tin is less reactive than Zn.

12. Anhydrous calcium chloride is deliquescent in nature which absorbs moisture and forms a solution. Thus, $CaCl_2$ solution formed does not allow the dust particles and other suspended particulate matter to rise.

13. If excess of carbon dioxide is passed through lime water, it becomes transparent due to the formation of $Ca(HCO_3)_2$.

14. O_2 is collected by downward displacement of water and CO_2 is collected by upward displacement of air.

15. A can be SO_2 or NO_2 because they pollute air and cause acid rain to pollute water. B can be CO_2 since it does not cause acid rain. Its presence in water is essential for aquatic plants. It is considered as an air pollutant as it results in global warming.

Test 2

1. Holes are created in the ozone layer due to the attack of CFCs which are released from refrigerators and air conditioners.

2. X is an acidic gas since it changes blue litmus to red. Y is neutral since it does not respond to either red or blue litmus test.

3. Burning of meteorites takes place in the mesosphere, aeroplanes fly in the stratosphere, reflection of radio waves takes place in the ionosphere. Magnetosphere absorbs ionising radiations and most of the weather phenomena occur in the troposphere.

4. A → d Tin coated iron articles are used to preserve edible food materials.

B → c Enamelling is done to washbasins and ovens to prevent rust and provide smooth appearance.

C → b Engine parts are periodically coated with oil grease to have durability.

D → a Iron trunks and sanitary pipes are coated with zinc to prevent corrosion.

5. Magnesium catches fire when it reacts with oxygen and continues to burn in an atmosphere of N_2.

6. (i) P_2O_5 on dissolution in water gives phosphoric acid.

(ii) CO_2 is a non-supporter of combustion, but allows burning of active metals.

(iii) Oxygen turns alkaline pyrogallol brown.

(iv) SO_2 when passed through potassium dichromate turns the solution from orange to green.

7. As we move to higher altitudes, fountain pen ink leaks out because of lower atmospheric pressure. The pressure built up in the ink pen is more when compared to the outside pressure. Hence, it leaks out.

8. Many fuel gases contain hydrogen as a major constituent because of its high calorific value.

9. Due to acid rain, sculptures and monuments get damaged and also water bodies like, rivers, oceans get acidified. This acidified water affects the aquatic life which leads to death and decay of aquatic animals.

10. Sulphur is the non-metal which burns with a blue-coloured flame. SO_2 turns orange-coloured $K_2Cr_2O_7$ green due to the formation of $Cr_2(SO_4)_3$.

11. Aluminium, when exposed to the atmosphere, reacts with air, i.e., O_2 and forms Al_2O_3. This layer of Al_2O_3 (alumina) makes aluminium passive.

12. The usage of refrigerators and air conditioners releases chlorofluoro carbons which don't contribute to SPM.

13. When carbon dioxide gas is passed through lime water, it gives a white precipitate due to the formation of calcium carbonate which is insoluble in water.

14. The gas is highly soluble in water. Hence, it cannot be collected by the downward displacement of water. It is collected by the upward displacement of air as it is heavier than air.

15. CO_2 and water vapour are greenhouse gases present in the atmosphere. Their percentage increase in atmosphere increases the temperature of the earth and contributes to global warming. These are the major contributors to global warming.

Chapter 5

Water

Remember

Before beginning this chapter, you should be able to:

- have knowledge of layers of the earth.
- have basic concept of hydrosphere.

Key Ideas

After completing this chapter, you should be able to:

- know about the different sources, forms of water and importance of water.
- explain the physical properties and uses of water, and depletion of water table.
- learn the different types of water (hard and soft), causes of hardness and methods of removal of different kinds of hardness.
- understand causes of water pollution and the process of purification of water.
- study about the volumetric composition of water by means of electrolysis.

INTRODUCTION

The uniqueness of the planet earth in the universe lies in it being inhabited by a wide variety of flora and fauna. Water is one of the most important abundant resources which is responsible for sustaining life on this planet alone.

The significance of water in the sustenance of the biosphere dates back to the evolution of aquatic life in oceans millions of years ago. It is believed that complex chemical reactions between the minerals in sea water took place in the presence of sunlight which resulted in the origin of unicellular organisms such as bacteria and fungi. The role of water is not confined to the evolution of life alone. Water is the essential constituent required for photosynthesis which is the only process that supports life. Being the major component in all living cells, water plays the most crucial role in all biological processes.

Distribution of water on earth's surface

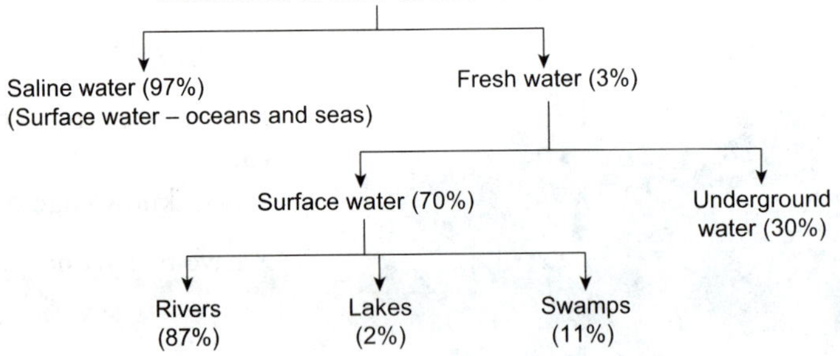

PHYSICAL PROPERTIES OF WATER

Water possesses certain significant physical properties which are responsible for making it a vital liquid.

1. **Nature:** It is a colourless, odourless and tasteless liquid.
2. **Freezing point:** Water freezes to ice at 0°C under normal atmospheric pressure, i.e., 76 cm of Hg.
3. **Boiling point:** Water gets converted into steam at 100°C under normal atmospheric pressure, i.e., 76 cm of Hg. The boiling point of water is proportional to external pressure.
4. **Density:** In general, solids possess greater densities than liquids. Therefore, for such liquids, the density decreases with an increase in temperature as the liquids expand on heating. However, in the case of water, it is different. Ice has lower density than water at the same temperature. As ice gets converted to water at 0°C its density increases and this trend continues till the temperature rises to 4°C. Beyond 4°C, water shows normal trend of decrease in density with increase in temperature. Therefore, density of water is maximum at 4°C (1 g/cc). From 4°C to 0°C, water undergoes expansion while all other liquids undergo contraction. Since this trend is opposite to the normal trend, this is called **anomalous expansion of water**.

TABLE 5.1 Density of water at different temperatures

Temperature	Density
0°C	0.91 g/cc (ice)
0°C	0.97 g/cc (water)
4°C	1 g/cc
>4°C	<1 g/cc

5. **Specific heat:** The amount of heat energy required to raise the temperature of unit mass of a substance through 1°C is called specific heat capacity of that substance. Specific heat capacity of pure water is 1 calorie/g/°C. Among all substances, water has the highest specific heat capacity.
6. **Latent heat:** When a substance changes from one state to another, certain amount of heat is required to overcome the intermolecular forces of attraction. For instance, when ice of certain mass at 0°C is converted into water of same mass at 0°C, the heat supplied brings about

conversion of state without change in temperature. This is stored as potential energy in the water molecules and is called latent heat of fusion and is equal to 80 cal/g.

Similarly, the conversion of 1 g of water at to steam at 100°C requires 540 cal heat and this is called **latent heat of vaporisation**.

At any temperature between 0°C and 100°C, conversion of water to water vapour takes place rather slowly. This process is called evaporation. The rate of evaporation increases with increase in temperature.

7. **Conductivity:** Pure water is a poor conductor of electricity. However, the presence of dissolved salts renders normal water a good conductor.
8. **Solvent property:** Water can dissolve most of the substances in it. Hence, it is known as **universal solvent**.

FORMS OF WATER

Water is the most abundant component on the earth's surface. In fact from 70 to 75 per cent of the earth's surface is occupied by water. Although the term 'water' is mostly relevant for liquid state, the occurrence of this substance in other states in nature is also considerable.

In very cold regions of world such as the Arctic circle and the Antarctica, water occurs in the form of snow or ice. High mountains also contain water in the form of ice due to very low temperatures prevailing there. Water exists as water vapour in the atmospheric air. Water vapour is present in a condensed form in clouds, fog, mist, etc. Liquid state is the most common state of water. Basically, water is available in two forms namely surface water and underground water.

SOURCES OF WATER

1. **Surface water:** It is the water present on the surface of the earth. The water present in oceans, seas, rivers, springs, etc., comes under surface water.

Oceans and seas: These are the largest sources of water. Almost 97 per cent of water available on the earth is present in oceans and seas. This water contains huge amount of dissolved salts in addition to other impurities. Due to the presence of high proportion of dissolved salts (especially 2.5 per cent common salt) which makes the water salty and renders the water unfit for usage and consumption, this called **saline water**. The salinity of sea water can be removed by the process of **desalination**.

Desalination: The process of removal of dissolved salts from the sea water is known as desalination. Desalination is carried out by reverse osmosis technique. In this method, sea water which contains dissolved salts is taken in the inner vessel and pure water is taken in the outer vessel. These two vessels are separated by a semipermeable membrane. Pressure greater than osmatic pressure is applied on the sea water with the help of a piston, as a result of which pure water from the inner vessel passes into the outer vessel, leaving behind the dissolved salts. Hence, desalination of sea water takes place.

Rivers and lakes: Rain water from higher altitudes flows down the hills and forms rivers. Apart from this rain water, water formed by melting of ice and snow flows into the rivers. Since water takes a long course and flows over land surface from higher altitudes, it contains various impurities and soluble salts. As the proportion of dissolved salts is less as compared to saline water, it is considered as fresh water and is available for consumption for various activities.

Rain water: It is the main source of natural water. Since it directly comes from the atmosphere, it is the purest form of water. The initial showers of rain water contain dust particles and other microorganisms. The later part of rain water is devoid of any impurities. Therefore, it is considered as the purest form of natural water.

2. **Underground water**: A part of the rain water reaches the water bodies such as rivers and lakes. A part of the rain water seeps into the soil and reaches the bottom layers. The rain water as well as some water from river water seeps through the soil and fills the empty spaces and cracks in the deep layers of the earth. This phenomenon is called **infiltration** and the water is called underground water. As water infiltrates through the open spaces in the soil, it first passes through the zone of aeration where the soil is unsaturated. At greater depths, known as recharge area, water fills in more spaces until the zone of saturation is reached. At this level, the ground water pressure is equal to the atmospheric pressure and this is known as water table. Below the water table, there is an underground layer of water bearing permeable rock or unconsolidated materials from which ground water can be usefully extracted using a water well. This zone is called **aquifer**.

   The underground water during seepage gets filtered through various layers of soil, and hence, it does not contain suspended impurities.

   Water in the water tube or aquifer can be drawn out by using hand pumps or tube wells. This water is called well water and it contains soluble impurities. The depth of well varies from place to place.

   Ground water while travelling through a network of cracks and fissures emerges from below the surface of the earth. At this point, the aquifer surface meets the ground surface and water flows onto the surface of the earth with high pressure. This spring water is devoid of suspended impurities but contains dissolved salts and minerals. Due to the presence of rare minerals, it is used for medicinal purposes.

USES OF WATER

1. Essential constituent for the reaction of photosynthesis which is the process providing support for life as well as making food for plants and animals.
2. Medium for all biochemical processes taking place in plants and animals.
3. Medium for transport of nutrients and minerals to the living cells in living beings.
4. Exchange of gases through blood by the process of respiration.
5. Germination of seeds.
6. Growth of plants.
7. Regulation of body temperature.

WATER SCARCITY

Scanty rainfall and drought conditions may lead to water scarcity. This in addition to reducing availability of water for human consumption, affects vegetation very badly. Water shortage results in wilting and drying up of plants and finally in the death of the vegetation. As rivers dry up in summer season, water is stored by constructing dams. This water is used for agricultural activities and also for the generation of hydel power.

WATER CYCLE

Water which is so essential in so many ways derives its major source from rains. Water is added to the atmosphere in the form of water vapour by various processes. The water vapour so added to the atmosphere, being lighter than air, reaches the upper atmosphere. There, due to lower temperature, it gets condensed to form tiny water droplets which remain suspended in air. These small water droplets associated with the dust particles form larger aggregates called **micelles**.

They appear as clouds and when the quantity of water in the clouds increases, it causes rain. This rain water can form various types of water bodies in the form of surface water and can help recharge ground water level. This natural process of adding water to atmospheric air and its subsequent condensation to rain water is called **water cycle**. Thus, water cycle builds a natural balance in the ecosystem.

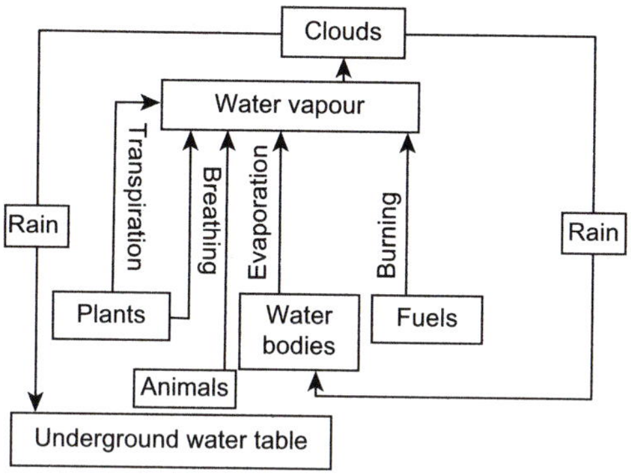

FIGURE 5.1 Water cycle

Depletion of Water Table

Sea water, which is the major proportion of water on the earth's surface, is saline and not suitable for consumption. Only river water and underground water making up to 0.01 per cent are sources of water. River water being limited and rivers drying up during the summer season, the dependence on underground water is inevitable.

The underground water is drawn out by using various devices such as hand pumps and tube wells. This water is again replenished by means of seepage of water into the ground during rainy season. When there is an imbalance between underground water and its replenishment, the level of water table may go down. This is called **depletion of water table**. This is more common in times of scanty rainfall.

Causes of Depletion

1. **Scanty rainfall or draught conditions:** As long as there is balance between the usage of underground water and its replenishment by rainfall, there is no question of depletion. But during long periods of scanty rainfall or in the absence of rainfall, this balance is disturbed. Usage of underground water continues more or less at the same level while its replenishment becomes very less or almost nil. This results in the depletion of water table.
2. **Deforestation:** Big trees in dense forests bind the soil particles through the roots. As a result, they help in reducing the water flow and help in the seepage of rain water through the soil. Deforestation results in reduction of forest area which allows the rain water flow rapidly into large water bodies. Seepage becomes less thereby leading to imbalance in the replenishment of underground water. Moreover, trees are one of the sources of water vapour which condenses to form clouds and results in rain.
3. **Increasing population:** Increasing population forces increased demand for construction. This leads to depletion of water table by decreasing the area of seepage and increasing the demand for water.
4. **Increasing industrialisation:** Every industry requires water for usage in various ways. Increase in number of industries leads to tremendous increase in usage of water which does not match with the rate of replenishment. Thus, it also eventually leads to the depletion of water table.

5. **Increased agricultural activities**: Due to erratic rainfall, the canal irrigation becomes rather impractical in certain areas. As a result, agriculture mostly depends on ground water for irrigation. The increased demand for agriculture results in the depletion of water table.

The above activities in the long run lead to tremendous depletion of ground water level and eventually to acute shortage of water.

Water being the most precious and essential natural resource, and the percentage of available water being less, it should be used judiciously. Several steps should be taken in day-to-day life both by the citizens as well as authorities for preventing depletion of water table.

Water Management

In view of the lesser abundance of available fresh water and the various natural factors affecting water cycle, water is considered as an invaluable natural resource. Therefore, water management is an important area to be dealt with.

Water management as such involves two major aspects. One is prevention of wastage of water by providing water supply through pipes without leakage. At the individual level, minimising the wastage in daily activities is an important step. Another aspect is utilising the rain water to recharge the ground water. This is called **water harvesting** or **rain water harvesting**. This can be done by installing water harvesting systems in the households.

Adoption of drip irrigation system in agriculture can also effectively help to minimise the usage of water in agricultural practices. In drip irrigation, water is supplied to the plants using narrow tubings which deliver water directly at the base of the plant. Irrigation by closed pipes instead of open canals can be of help in this direction since it reduces the loss of water by evaporation.

In regions with scanty rainfall, a substance called cetyl alcohol is sprayed on the water in reservoirs meant for storage of large quantities of water. This reduces rate of evaporation of water.

Growing some plants under plastic covers also minimises water loss as the water vapour condenses on the inner surface of plastic cover and is given back to soil.

Hard and Soft Water

Underground water is the main source of water used for various purposes. Since this water remains in the bottom layers of earth's crust for a long time, it contains many dissolved salts. Depending on the type of dissolved salts, water can be classified into two types:

Soft water: Water which can freely form lather with soap is called **soft water**.

Hard water: Water which does not form lather with soap and forms a precipitate or scum is called **hard water**.

Causes of Hardness

If the sample of water contains calcium and magnesium bicarbonates, chlorides and sulphates; they do not allow the formation of much lather. Therefore, these salts cause hardness of water.

Disadvantages of Hardness

Due to the presence of various dissolved salts in water, hard water is not suitable for use for various purposes.

1. Not suitable for drinking.
2. Due to very poor lather formation, it is not effective for washing, cleaning and bathing purposes.

3. Scale formation in boilers results in loss of heat and wastage of fuels in industries.
4. Not effective for dyeing of fabrics due to improper fixing of dye to the fabric.

Types of Hardness and Removal of Hardness

Among these salts, presence of bicarbonates in water results due to the dissolution of CO_2 gas in rain water. This water flows across a wide area with different types of rocks. Calcium carbonate and magnesium carbonate present in the rocks react with the water containing CO_2 thereby forming the corresponding bicarbonates. Bicarbonates being highly soluble in water impart hardness to water.

The hardness caused by these bicarbonate salts is called **temporary hardness**.

$$CaCO_3 + H_2O + CO_2 \rightarrow Ca(HCO_3)_2$$

$$MgCO_3 + H_2O + CO_2 \rightarrow Mg(HCO_3)_2$$

Removal of Temporary Hardness

1. **Boiling:** However, when water is subjected to boiling, these bicarbonates get converted into insoluble carbonates and can be removed by filtration. Since the hardness is removed by simple boiling, it is called temporary hardness and the water is called **temporary hard water**.

$$Ca(HCO_3)_2 \xrightarrow{\Delta} CaCO_3 + H_2O + CO_2$$

$$Mg(HCO_3)_2 \xrightarrow{\Delta} MgCO_3 + H_2O + CO_2$$

2. **Clark's method:** This method involves the addition of slaked lime to water either in solid or liquid form. This also results in conversion of soluble bicarbonates to insoluble carbonates. On filtration, soft water is obtained.

$$Ca(HCO_3)_2 + Ca(OH)_2 \longrightarrow 2CaCO_3 \downarrow + 2H_2O$$

$$Mg(HCO_3)_2 + Ca(OH)_2 \longrightarrow CaCO_3 \downarrow MgCO_3 \downarrow + 2H_2O$$

If water dissolves other minerals in it, such as chlorides and sulphates; they also cause hardness of water. However, these salts cannot be removed by boiling. Hence, the hardness imparted by these salts is called **permanent hardness**.

Removal of Permanent Hardness

There are various methods adopted to remove permanent hardness.

1. **Addition of washing soda (sodium carbonate):** This method can remove both temporary as well as permanent hardness from water.

   Bicarbonate salts as well as sulphate- and chloride salts undergo double decomposition reaction with sodium carbonate. In these reactions, the insoluble carbonates formed get precipitated and are removed by filtration.

$$Ca(HCO_3)_2 + Na_2CO_3 \rightarrow CaCO_3\downarrow + 2NaHCO_3$$

$$MgSO_4 + Na_2CO_3 \rightarrow MgCO_3\downarrow + Na_2SO_4$$

$$CaCl_2 + Na_2CO_3 \rightarrow CaCO_3\downarrow + 2NaCl$$

2. **Permutit process:** Hydrated sodium aluminium silicate ($Na_2Al_2Si_2O_8.xH_2O$) is called **permutit**. The sample of hard water is allowed to pass through a cylindrical tube filled with permutit. As the water passes through permutit, the calcium and magnesium ions of hard water are replaced by sodium ions. Sodium ions do not cause any precipitation with soap and do not hinder the formation of lather. Thus, the water is said to be softened.

After some time, the entire permutit contains Ca^{+2} and Mg^{+2} in place of Na^+ ions. This is called exhausted permutit. This can be regenerated by sending sodium chloride solution through exhausted permutit.

3. **Ion exchange method:** As the name indicates, this method involves exchange of hardness-causing ions by other ions. In this method, the apparatus consists of two cylindrical tubes A and B known as cation exchanger and anion exchanger, respectively.

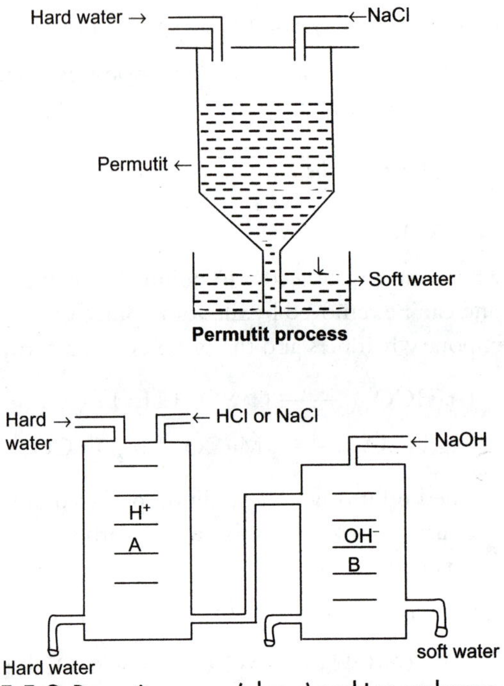

FIGURE 5.2 Permutit process (above) and Ion exchange process

These ion exchangers contain porous solids which do not dissolve in water but swell in water. These solids are called **ion exchange resins**. They contain mobile groups or ions which can be exchanged with other ions. The resin-containing groups or ions capable of exchanging cations are called **cation exchange resins** and the resin-containing groups or ions capable of exchanging anions are called **anion exchange resins**. H-cation exchanger results in exchange of Ca^{+2} or Mg^{+2} ions by H^+ ions. **Na-cation exchanger** results in exchange of Ca^{+2} or Mg^{+2} ions by Na^+ ions. Hard water is first passed through tube A where all the positive ions are displaced by H^+ or Na^+ ions depending on the type of resin used. The water is then passed through tube B where all the negative ions are displaced by OH^- ions present in anion exchange resin. Thus, the water coming out of ion exchange apparatus now contains only pure water. This is also called **demineralised water**. When the cation exchanger and anion exchanger get depleted of H^+ and OH^- ions, the ion exchange resins are regenerated by passing HCl solution or NaCl solution as the case may be and NaOH solution through the tubes A and B, respectively.

$$MgCl_2 + 2H^+ \text{ Resin} \rightarrow Mg - \text{Resin} + 2HCl$$

$$RNH_3OH + Cl^- \rightarrow RNH_3Cl + OH^-$$

$$H^+ + OH^- \rightarrow H_2O$$

Once hardness is removed, the water can be used for various day-to-day activities, industries, etc. In addition to these naturally added components, i.e., dissolved salts, many other components are added to the water by human activities. These components are called **pollutants**.

EXAMPLE

State and explain the anomalous expansion of water.

SOLUTION

In general, solids possess greater densities than liquids. Therefore, for such liquids, the density decreases with an increase in temperature as the liquids expand on heating. However, in the case of water it is different. Ice has lower density than water at the same temperature. As ice gets converted to water at 0°C, its density increases and this trend continues till the temperature rises to 4°C. Beyond 4°C, water shows normal trend of decrease in density with increase in temperature. Therefore, density of water is maximum at 4°C (1 g/cc). From 4 to 0°C, water undergoes expansion while all other liquids undergo contraction. Since this trend is opposite to the normal trend, this is called **anomalous expansion of water**.

EXAMPLE

What is meant by infiltration? Explain the role of infiltration in water management.

SOLUTION

A part of the rain water reaches the water bodies such as rivers and lakes. A part of the rain water seeps into the soil and reaches the bottom layers. The rain water as well as some water from the river seeps through the soil and fills the empty spaces and cracks deep below the ground. This phenomenon is called infiltration. The infiltration of water helps in the recharge of water table.

EXAMPLE

The electronic configurations of two metals X and Y are 2, 8, 2 and 2, 8, 8, 2, respectively. What is the effect of the presence of hydrogen carbonate and sulphates of these metals on water?

SOLUTION

Elements X and Y are Mg and Ca, respectively. Hydrogen carbonates or bicarbonates of Ca and Mg impart temporary hardness while sulphates of Ca and Mg impart permanent hardness to water.

EXAMPLE

In permutit process, what is the role of the addition of the concentrated solution of the solid compound obtained due to the reaction between NaOH with HCl? Explain.

SOLUTION

In permutit process, sodium permutit that is $Na_2Al_2Si_2O_8.xH_2O$ is used. When hard water is treated with sodium permutit, Ca^{+2} and Mg^{+2} ions are exchanged with the Na^+ ions present in the permutit. After long use, most of the sodium ions present in permutit are replaced with calcium and magnesium ions, and hence, it becomes ineffective. Sodium hydroxide reacts with HCl to form NaCl. Concentrated NaCl solution is added to regenerate the permutit.

EXAMPLE

Jini took a container completely filled with water with some ice cubes floating on it. He observed the container expecting that water to overflow after the ice melted. Can you predict what he would finally see?

SOLUTION

The water would not overflow after the melting of the ice cubes because the volume of water obtained from the ice cubes is equal to the volume of water displaced by the ice cubes. The container would be filled up to the brim.

EXAMPLE

Why is water called universal solvent?

SOLUTION

Water can dissolve most of the substances in it, and hence, it is known as universal solvent.

WATER POLLUTION

TABLE 5.2 Cause and control of water pollution

Causes for water pollution	Types of pollutants	Effects on environment	Control measures
(a) Defecation by living beings (excreta of animals and humans)	Microorganisms (tapeworm, round worm, hookworm, amoeba, etc.)	Water-borne diseases at times lead to epidemics	(a) Usage of pit latrines (b) Conversion of cow dung into biogas and the left over slurry can be used as manure
(b) Industrial operation	Poisonous metallic salts (mercury, cadmium, chromium and arsenic)	Affects aquatic organisms	Removal of metallic salts by chemical treatment of water before discharging it
(c) Agricultural activities	Residual amount of artificial fertilisers, pesticides, insecticides and fungicides	Pollutes underground water	(a) Usage of natural manure in place of artificial fertilisers (b) Usage of natural pesticides (c) Avoiding spraying of chemicals in rainy season (d) Usage of drip irrigation
(d) Thermal pollution (steel and thermal power plants) Atomic reactors	Hot water	Favourable for growth of algae leading to decrease in dissolved O_2 which affects aquatic life	(a) Cooling and recycling of water in the industry (b) Cooling by cooling towers before discharging it into water bodies

Potable Water or Drinking Water

Many steps are taken by authorities to control water pollution. One such important programme is 'Ganga Action Plan.' This plan envisages the enforcement of laws to force the industries to set up water treatment plants. In spite of taking all the control measures, water from natural sources still contains some suspended impurities and microorganisms. Although it can be used safely for other activities, it cannot be used for the purpose of drinking or human consumption. Therefore, it is necessary to subject this water to purification.

The purification of water basically involves two steps.

1. Removal of suspended impurities and
2. Removal of microorganisms.

1. **Removal of suspended impurities**

   (a) **Sedimentation**: River water is pumped into settling tanks and allowed to settle without any disturbance. The suspended impurities being larger particles settle down on standing.

   (b) **Addition of chemicals**: The above process makes water only semi-clear. This is passed into a mixing chamber where alum and lime are added to the water and it is agitated with huge mechanical blades. Later on, this water is pumped into a sedimentation tank. Alum hastens the process of precipitation of suspended impurities. During the precipitation process, some bacteria also get trapped and removed.

   (c) **Filtration through sand and gravel**: The clear water obtained is pumped into a sand and gravel filter. This removes all remaining traces of suspended impurities.

2. **Removal of microorganisms**

   Addition of bleaching powder releases chlorine which helps in killing the harmful bacteria and the pure water obtained is stored in storage tanks.

   The water so obtained at the end of the above purification process is devoid of suspended impurities and harmful bacteria. This water is fit for drinking and is called **potable water**. Potable water contains dissolved gases, such as O_2 and CO_2, and also some dissolved salts especially common salt. These salts play an important role in the metabolic activities. They also provide taste to water. Therefore, this water is very much suitable for human consumption.

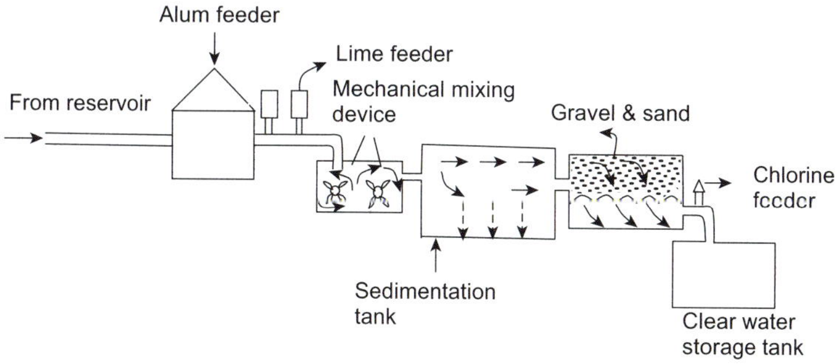

FIGURE 5.3 Purification of water

Distilled Water

Since most of the chemical reactions take place in water, laboratory work requires pure water for carrying out various tests and reactions. For this purpose, the dissolved salts should be removed from water to get pure water. This is done by subjecting water for purification by the process of distillation.

Distillation Using Liebig Condenser

The water to be purified is taken in a distillation flask and subjected to boiling. The steam formed as a result of boiling when passed through Liebig condenser condenses due to cooling and gets collected in the receiver flask. The dissolved salts are left behind in the distillation flask.

Distillation Using Continuous Water-Still

The sample of water is warmed by condensing steam prior to subjecting the water for boiling in the still. The advantage of the process is that it requires usage of lesser amount of fuel for boiling the water in the still as it had already been heated by condensing steam.

Uses of Distilled Water

1. Used exclusively for chemical work in laboratories.
2. It has a bland and insipid taste and in fact considered as tasteless since the mineral salts and dissolved gases only impart taste and freshness to water.
3. It is not a substitute for potable water. Although it is a pure form of water, it does not contain the required mineral salts and dissolved gases. The nutrients from body fluids get dissolved in this water and are sent out, thus, leading to deficiency of vital nutrients.
4. The saline water from seas can be subjected to distillation and to the water so obtained, little amount of common salt is added for making it suitable for use as potable water.

Determination of volumetric composition of water by the process of electrolysis: Decomposition is a chemical change and can be brought about by using various forms of energy like light, electricity, etc. The decomposition of a substance by passing electricity is called **electrolysis**.

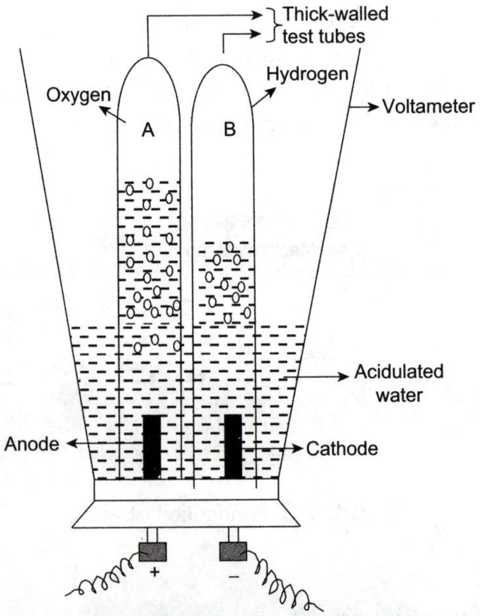

FIGURE 5.4 Electrolysis of water

Principle: Since pure water is a poor conductor of electricity, little amount of sulphuric acid is added to water. This acidulated water on passage of electricity decomposes to give hydrogen and oxygen in 2 : 1 volume ratio.

$$2H_2O \rightarrow 2H_2 + O_2$$

Process: The acidulated water is filled in two test tubes without any air bubbles. The two test tubes are inverted over two iron nails connected to positive and negative terminals of a battery. These iron nails act as electrodes. The iron nail connected to positive terminal of the battery is called **anode** and the one connected to negative terminal is called **cathode**.

On passing electricity for some time, two gases are collected in the two test tubes. Corresponding decrease in the volumes of solution in the test tubes indicates the volumes of gases formed. The gases in the respective test tubes can be identified by introducing a burning splinter. In test tube A the splinter burns brilliantly. In test tube B, the splinter is put off with a pop sound and the gas burns with pale blue flame. The former is identified as oxygen, whereas the latter as hydrogen. The volumes of oxygen and hydrogen are found to be in 1 : 2 ratio.

When water is formed from its constituent elements also, the combination reaction reveals that the volumetric composition of water corresponds to hydrogen and oxygen in 2 : 1 ratio.

$$2H_2 + O_2 \rightarrow 2H_2O$$

EXAMPLE

What is the advantage of distillation by using continuous water still over distillation by using Liebig condenser?

SOLUTION

The advantage of distillation by using continuous water still is that it requires usage of lesser amount of fuel for boiling the water in the still as it had already been heated by condensing steam.

EXAMPLE

Why is distilled water not suitable for drinking?

SOLUTION

Distilled water does not contain the required mineral salts and dissolved gases. The nutrients from body fluids get dissolved in this water and are sent out, thus, leading to deficiency of vital nutrients. Distilled water has a bland and insipid taste and is in fact considered as tasteless. Hence, distilled water is not suitable for drinking.

POINTS TO REMEMBER

- Among the different sources of water, rain water is considered to be the purest.
- Although distilled water is a pure form of water, we use potable water for drinking.
- Specific heat of water which is 1 cal $g^{-1}{}^{\circ}C^{-1}$ is the highest for any known substance. The latent heat of fusion is 80 cal/g/°C and latent heat of vaporisation is 540 cal/g/°C.

- Water shows anomalous expansion. Density of water is maximum at 4°C which is equal to 1 g/cc.
- The boiling point is directly proportional to external pressure.
- Due to the presence of calcium and magnesium salts, water does not form lather and such water is called hard water.
- Hardness in water may be classified into temporary hardness and permanent hardness. Temporary hardness can be removed by boiling or Clark's method and permanent hardness is removed by permutit process or ion exchange method.
- Pollution of water is caused by defecation by living beings, industries, agricultural operations, thermal power plants, etc.
- Water pollution can be controlled by converting human excreta, cow dung, etc., into biogas, use of natural fertilisers, pesticides and by adopting drip irrigation technique.
- The volumetric composition of water is hydrogen and oxygen in the ratio of 2 : 1.

TEST YOUR CONCEPTS

Directions for questions from 1 to 10
Fill in the blanks.

1. ________ per cent of the total water available on the earth is saline water.
2. Drinking water mainly contains small amounts of salts of _______, ______and _______ metals.
3. Carbon dioxide on dissolution in water forms ________.
4. 1 g of water at 100°C on heating forms 1g of steam at the same temperature by absorbing _________ cal of energy.
5. 1g of pure water at 4°C occupies a volume of ________.
6. Water containing calcium bicarbonate produces _________ with soap.
7. _________ volume(s) of hydrogen and _________ volume(s) of oxygen under similar conditions of temperature and pressure combine to give two volumes of water.
8. _________ hardness is removed by Clark's method.
9. _________ is added to prevent the evaporation of water.
10. Watering plants by using narrow tubes is called ________.

Directions for questions from 11 to 25
For each of the questions, four choices have been provided. Select the correct alternative.

11. Which of the following chemicals lead to water pollution?
    (a) fertilisers (b) insecticides
    (c) pesticides (d) all the above
12. Freezing point of water is 0°C at ______.
    (a) 76 cm of Hg (b) 760 cm of Hg
    (c) 76 mm of Hg (d) 7.6 mm of Hg
13. The salts of which among the pairs of metals cause greater water pollution?
    (a) calcium, aluminium (b) mercury, cadmium
    (c) calcium, cadmium (d) calcium, mercury
14. Identify the set of pure and impure water among the following.
    (a) rain water and sea water
    (b) rain water and distilled water
    (c) sea water and river water
    (d) river water and ground water
15. Chemical treatment of surface water can be carried out by the addition of
    (a) bleaching power (b) alum
    (c) oxygen (d) carbon dioxide
16. Density of water is less than 1g/c.c ________.
    (a) at 4°C (b) above 4°C
    (c) below 4°C (d) both (b) and (c)
17. Latent heat of vaporisation of water is ________.
    (a) 270 cal/g (b) 540 cal/g
    (c) 80 cal/g (d) 180 cal/g
18. Which of the following processes can be used to remove both temporary and permanent hardness?
    (a) Clark's method
    (b) addition of washing soda
    (c) boiling
    (d) all the above
19. Which of the following terms is not related to underground water?
    (a) infiltration (b) aquifer
    (c) water table (d) potable water
20. Which of the following water is suitable for drinking?
    (a) soft water
    (b) demineralised water
    (c) potable water
    (d) distilled water
21. When the water gets converted into ice, its volume
    (a) decreases
    (b) increases
    (c) remains the same
    (d) initially decreases and then increases
22. Which of the following dissolved gases are present in drinking water?
    (a) oxygen and nitric oxide
    (b) oxygen and carbon monoxide
    (c) oxygen and carbon dioxide
    (d) oxygen and nitrogen

23. Which of the following salts is responsible for temporary and permanent hardness of water, respectively?
(a) $Mg(HCO_3)_2$ and Na_2SO_4
(b) $Ca(HCO_3)_2$ and $MgCl_2$
(c) $MgSO_4$ and $Mg(HCO_3)_2$
(d) $CaCl_2$ and $CaSO_4$

24. Arrange the following terms in sequence from source to supply.
(1) chlorination
(2) sedimentation
(3) filtration
(4) storage water tank
(5) lakes
(a) 53142 (b) 13425
(c) 52314 (d) 23145

25. Arrange the terms in sequence of water cycle starting from water bodies.
(1) precipitation
(2) condensation
(3) water vapour
(4) infiltration
(5) Ground water table
(a) 54231 (b) 32145
(c) 35412 (d) 52314

Directions for questions from 26 to 28

Match the entries given in Column A with the appropriate ones in Column B. More than one suitable content of Column B can also be considered correct for an entry in Column A and some content given in Column B may not be the appropriate one for any entry in Column A.

26.

Column A	Column B
A. Hard water	( ) a. Reverse osmosis
B. Soft water	( ) b. Rain water
C. Desalinated water	( ) c. Presence of Ca^{+2} and Mg^{2+}
D. Lake water	( ) d. Scum formation
	( ) e. Lather formation

27.

Column A	Column B
A. Potable water	( ) a. Electrolysis
B. Distilled water	( ) b. Irrigation
C. Acidulated water	( ) c. Pure water
D. River water	( ) d. Laboratory purpose
	( ) e. Human consumption

28.

Column A	Column B
A. Sedimentation	( ) a. Softened water
B. Filtration	( ) b. Addition of alum
C. Chlorination	( ) c. Removal of suspended impurities
D. Permutit	( ) d. Purification of drinking water

29. Why does river water contains lot of impurities?

30. What is the volumetric composition of water?

Very Short Answer Type Questions

31. What is the role of water in the evolution of aquatic life?

32. Name the most vital process where water takes part in the reaction in which one of the products is a supporter of life.

33. What is an aquifer?

34. Underground water does not contain suspended impurities. Why?

35. What is meant by rain water harvesting?

36. Distinguish between hard water and soft water.

37. Compare the density of water at 4°C with that below and above 4°C.

38. What are the values of latent heat of fusion of ice and latent heat of vaporisation for water?

39. Why is spring water used for medicinal purpose?

40. Name the substance added to water stored in reservoirs in regions with scanty rainfall. Give reason.

41. What is permutit? Give its formula?

42. Why is ordinary water a good conductor of electricity?

43. How are springs formed?

44. What is desalination?

45. Define electrolysis.

46. What is depletion of water table?

47. Distinguish between temporary hardness and permanent hardness.

48. Why is rain water considered the purest form of natural water?

Short Answer Type Questions

49. What is meant by infiltration? Explain the role of infiltration in water management.

50. Give some causes of depletion of water table.

51. Name some agricultural practices which help in water management.

52. How are bicarbonates added to water? How can they be removed by boiling? Give equations.

53. Explain the removal of temporary hardness of water by Clark's method. Give equations.

54. Give the characteristics of potable water.

55. Explain the role of alum in the purification of water.

56. Describe the various stages of purification of drinking water

57. Draw a neat labelled diagram for the electrolysis of water.

58. Explain the process by which both removal of temporary hardness and permanent hardness is possible.

59. Write a short note on water management.

60. Draw a neat line sketch of water cycle.

61. What is meant by depletion of water table? Explain in detail.

62. How can you harvest water in household?

63. Draw a line sketch of distribution of water on earth's surface.

64. List important uses of water.

65. Discuss five important physical properties of water.

66. Describe various sources of water.

Essay Type Questions

67. Explain the electrolysis of water to determine the volumetric composition of water with a diagram.

68. Give reasons for the following:
    (a) Sea water is unfit for usage and consumption
    (b) River water is considered as fresh water
    (c) Underground water does not contain suspended impurities.
    (d) Distilled water is used for laboratory processes.

69. Explain water cycle and its role in raising the ground water table.

70. What are the disadvantages of hard water? Explain the various methods of removal of permanent hardness of water with diagrams.

71. Discuss the various steps involved in making potable water.

72. (a) Mention the causes for water pollution.
    (b) Mention different types of pollutants.
    (c) Mention the control measures for water pollution.

73. (a) Explain the different causes of depletion of water table.
    (b) What steps should be taken in day-to-day life to prevent depletion of water table?

74. Mention suitable terms for the following statements:
    (a) Decrease in density of water on decreasing temperature from 4°C to 0°C which is opposite to the normal trend.
    (b) The underground layer of water bearing permeable rock present below the water table
    (c) The natural process of adding water to atmospheric air and its subsequent condensation to rain water.
    (d) Type of chemical reaction carried out when electricity is passed through water.

For Hints and Explanations, please visit:
www.pearsoned.co.in/IITFoundationSeries

CONCEPT APPLICATION

Level 1

Directions for questions from 1 to 10

State whether the following statements are true or false.

1. Underground water is free from suspended impurities.
2. Saline water contains large amounts of dissolved salts.
3. 1 g of pure water at 4°C occupies a volume of 0.97 cc.
4. Water containing calcium bicarbonate produces scum with soap.
5. Hard water is fit for dyeing clothes.
6. Cation exchange resin is regenerated by passing HCl solution.
7. Salt is added to pure water to increase its electrical conductivity.
8. Two volumes of hydrogen and one volume of oxygen compose one volume of water.
9. Permanent hardness is removed by Clark's method.
10. Specific heat of water is 1 cal/g/°C.

Directions for questions from 11 to 20

Fill in the blanks.

11. ______ and ______ are two types of underground water.
12. Hydrated sodium aluminium silicate is called ______.
13. The exhausted permutit contains ______ and ______ ions in place of Na^+ ions.
14. In sedimentation process, alum ______ the process of precipitation of suspended impurities.
15. Addition of ______ releases ______ which helps in killing the harmful bacteria present in water.
16. Addition of acid to water, ______ the conductivity.
17. In electrolysis of water ______ gas is liberated at cathode.
18. The change of water into water vapour at 30°C at slow rate is called ______.
19. ______ Small water droplets associated with dust particles to form larger aggregates called ______
20. Percentage of water in animals varies from ______ to ______.

Directions for questions from 21 to 37

For each of the questions, four choices have been provided. Select the correct alternative.

21. Which of the following processes is not involved in the purification of drinking water?
    (a) sedimentation (b) filtration
    (c) chlorination (d) distillation
22. Which of the following sources of water is not used for drinking purpose due to dissolved salts present in them?
    (a) rain water (b) sea water
    (c) spring water (d) river water
23. Density of water is maximum
    (a) at 0°C (b) at 4°C
    (c) above 4°C (d) below 4°C
24. Which of the following substances used in agriculture does not cause water pollution to a large extent?
    (a) manures (b) pesticides
    (c) fertilisers (d) insecticides
25. Metals present in permutit are
    (a) Na and K (b) Na and Al
    (c) Al and K (d) K and Al
26. Which among the following liquids has the highest specific heat?
    (a) petrol (b) mercury
    (c) oil (d) water
27. Which among the following types of water is a poor conductor of electricity?
    (a) saline water (b) tap water
    (c) river water (d) distilled water
28. Which of the following is not the control measure of water pollution?
    (a) usage of natural manure
    (b) usage of drip irrigation
    (c) usage of fertilisers
    (d) usage of natural pesticides
29. The pair of ions which does not cause hardness of water is
    (a) sulphate, chloride
    (b) bicarbonate, chloride
    (c) bicarbonate, sulphate
    (d) nitrate, phosphate

30. Regeneration of anion exchange resin is carried out by passing
(a) sodium hydroxide solution
(b) sodium chloride solution
(c) sodium bicarbonate solution
(d) all the above

31. Spring water usually contains
(a) dissolved salts (b) minerals
(c) suspended impurities (d) both 1 and 2

32. Which of the following water does not contain dissolved gases such as oxygen and carbon dioxide?
(a) potable water (b) distilled water
(c) saline water (d) soft water

33. Which of the following chemical substances can be used for removing permanent hardness of water by precipitation reaction?
(a) baking soda (b) washing soda
(c) alum (d) quick lime

34. When the temperature of water rises from 0°C to 10°C, the density of water
(a) decreases gradually
(b) decreases up to 4°C and then increases
(c) increases up to 4°C and then decreases
(d) increases gradually

35. The volumetric composition of water is
(a) 1 : 2 ratio of hydrogen and oxygen, respectively
(b) 1 : 8 ratio of hydrogen and oxygen, respectively
(c) 1 : 1 ratio of hydrogen and oxygen, respectively
(d) 2 : 1 ratio of hydrogen and oxygen, respectively

36. Arrange the process in sequence for the purification of drinking water
(1) lime feeder, alum feeder
(2) filtration
(3) mechanical mixing
(4) chlorination
(a) 1324 (b) 1243
(c) 2345 (d) 2341

37. Arrange statements in sequence of increasing order of heat content.
(1) ice of 10 g mass at 10°C
(2) water of 10 g mass at 10°C
(3) ice of 10 g mass at 40°C
(4) water of 10 g mass at 40°C
(5) steam of 10 g mass at 10°C
(a) 13245 (b) 54312
(c) 12534 (d) 12543

Directions for questions from 38 to 40

Match the entries given in Column A with the appropriate ones in Column B. More than one suitable content of Column B can also be considered correct for an entry in Column A and some content given in Column B may not be the appropriate one for any entry in Column A.

38.

Column A	Column B
A. Defecation by living beings	( ) a. Arsenic, mercury, etc.
B. Industrial operation	( ) b. Insecticides, fungicides, etc.
C. Thermal pollution	( ) c. Amoeba, fungicides, etc.
D. Agricultural activities	( ) d. Favourable for growth of algae

39.

Column A	Column B
A. Removal of hardness by boiling	( ) a. Resins
B. Clark's method	( ) b. Can remove temporary and permanent hardness
C. Addition of washing soda	( ) c. Decomposition reaction
D. Ion exchange method	( ) d. Addition of slaked lime

40.

Column A	Column B
A. Specific heat of water	( ) a. 250 cal/g°C
B. Latent heat of fusion of ice	( ) b. 0°C
C. Latent heat of vaporisation of water	( ) c. 100°C
D. Temperature at which melting of ice starts under normal atmospheric pressure	( ) d. 1 cal/g°C
E. Temperature at which boiling of water starts under normal atmospheric pressure	( ) e. 80 cal/g°C
	( ) f. 540 cal/g°C

Level 2

Directions for questions from 1 to 18

For each of the questions, four choices have been provided. Select the correct alternative.

1. What is the latent heat of fusion and latent heat of vaporisation of 10 g of ice and water, respectively?
   (a) 800 cal and 800 cal (b) 540 cal and 540 cal
   (c) 800 cal and 5400 cal (d) 540 cal and 800 cal

2. Which of the following properties of iceberg led to the sinking of the titanic ship?
   (a) heavy water currents
   (b) relatively higher density of ice
   (c) 9/10 of iceberg submerged below sea water and 1/10 above the sea water
   (d) none of the above

3. Which of the following processes is not involved in making surface water potable?
   (a) sedimentation (b) chlorination
   (c) filtration (d) distillation

4. The solvent water is used in car radiators. Which of the following properties of water is exploited?
   (a) high solubility (b) poor conductivity
   (c) maximum density (d) high specific heat

5. A divalent metal salt X, which contributes to hardness of water, combines with washing soda and forms an insoluble salt Y and common salt. Salt Y is also used for the laboratory preparation of CO_2. Identify X and Y, respectively.
   (a) $CaSO_4$, $CaCO_3$ (b) $MgSO_4$, $MgCO_3$
   (c) $CaCl_2$, $CaCO_3$ (d) $MgCl_2$, $MgCO_3$

6. The amount of heat energy required to increase the temperature of 20 g of water by 1°C is ______.
   (a) 10 cal (b) 20 cal
   (c) 15 cal (d) 2 cal

7. On passing electricity through acidulated water, the gaseous products obtained are collected in two separate test tubes A and B. The volume of the gas collected in test tube A is double the volume of gas collected in test tube B. Identify the two gases in test tubes A and B.
   (a) hydrogen and oxygen
   (b) oxygen and hydrogen
   (c) hydrogen peroxide and oxygen
   (d) oxygen and water vapour

8. Which of the following types of water is a bad conductor of electricity?
   (a) distilled water
   (b) water obtained in ion-exchange process
   (c) mineral water
   (d) both 1 and 2

9. The pressure cooker reduces the cooking time because of
   (a) increase in the boiling point of water
   (b) high specific heat of water
   (c) decrease in the boiling point of water
   (d) high latent heat of vaporisation

10. Wealthy but not healthy agriculture can be achieved by
    (a) introducing biogas plants
    (b) using natural pesticides
    (c) using drip irrigation
    (d) introducing artificial fertilisers

11. Two persons A and B got burns, one with boiling water at 100°C and other due to steam at 100°C, respectively. Which person is affected more?
    (a) A
    (b) B
    (c) both A and B to the same extent
    (d) none of the above

12. Aquatic animals are able to survive in water bodies like rivers and oceans during winter when the atmospheric temperature is –10°C. This is because
    (a) density of ice is less than water
    (b) density of water is minimum at 4°C
    (c) ice is a poor conductor of heat
    (d) both 1 and 3

13. Which of the following elements is used for the purification of surface water to make it free from germs so that it can be used for drinking purpose?
    (a) oxygen (b) fluorine
    (c) chlorine (d) carbon

14. Which of the following is the application of the highest specific heat capacity of water?
    (a) cooling of soft drink bottles in ice
    (b) cooling of earth's atmosphere after a hail storm
    (c) cool breeze from air coolers
    (d) bursting of water pipes during winter in cold countries

15. Water-containing salt X of a divalent metal, when treated with a compound Y gives much lather with soap. Insoluble compound $MgCO_3$ and salt of a monovalent metal sulphate are formed. Identify X and Y, respectively.
   (a) $CaCl_2$, Na_2CO_3
   (b) $MgSO_4$, Na_2CO_3
   (c) MgCl $NaHCO_3$
   (d) $CaSO_4$, $Mg(HCO_3)_2$

16. The amount of heat energy required to increase the temperature of X g of water by 10°C is found to be 15 cal. Calculate X.
   (a) 0.5 g (b) 15 g
   (c) 1.5 g (d) 7.5 g

17. The water obtained in ion-exchange process is
   (a) devoid of dissolved salts and minerals
   (b) devoid of only minerals
   (c) potable
   (d) called mineral water

18. Which among the following is based on the principle that boiling point decreases with decrease in atmosphere pressure?
   (a) Cooking food at sea level takes longer period of time.
   (b) Cooking food at higher altitudes takes longer period of time.
   (c) Working principle of air coolers.
   (d) Working principle of pressure cooker.

19. A group of students accompanied by their science teacher went trekking. After reaching the top of the hill, they felt hungry. The teacher told them that they would have to bear the hunger as cooking of food is slow and takes a longer time at the top of the hill. The students wanted to know the reason for the above fact. The teacher explained the reason for the above fact. Can you predict the explanation given by the teacher?

20. In a science exhibition, a student called Rishi performed the following experiment. She took two containers of the same dimensions and filled one of them with wax and the other one with ice. She placed both of them in a boiling water bath. What difference could she observe during the melting of these two substances? Also predict the reason for the observed difference.

21. Why is sea water not suitable for drinking? How is it made potable?

22. Hot water bags are generally used for reducing body pain. Can they also be used to keep us warm in winter? Explain.

23. Though ice is the solid form of water, it floats on the surface of water. Justify.

24. Calculate the amount of heat energy required to increase the temperature of 10 g of water through 12°C.

25. During summer vacations, Aryan went for a trip to Tirupathi in an old car along with his family. During the journey, the driver stopped the car beside a hotel and brought a can full of water and poured it into the radiator of the car. Aryan's son, Mouryan wanted to know the reason for pouring water into the radiator and asked his father who is a mechanical engineer by profession the reason. Aryan explained the role of water in the radiator of the car engine. Can you predict the explanation given by Aryan?

26. Calculate the ratio of the amount of heat energy lost during the conversion of 20 g of water at 0°C to ice at 0°C to heat energy gained during the conversion of same amount of water at 100°C to steam at 100°C.

27. Two open containers A and B are filled with water and in container A, cetyl alcohol is added. Both the containers are placed at 40°C. Predict the level of water in the two containers after some time and give reasons.

28. The electronic configurations of two metals X and Y are 2, 8, 2 and 2, 8, 8, 2, respectively. What is the effect of the presence of hydrogen carbonate and sulphates of these metals in water?

29. A group of students went for an excursion to Shimla in the month of November. A science teacher also accompanied them. There the temperature fell to −10°C. They took some ice cubes and subjected them to heating while keeping a thermometer in the container. The students monitored the readings of thermometer till the entire ice melted into water. All of them went to their science teacher and showed him their observations. He asked them to discuss among themselves and give a proper justification for the observations on the readings at various stages.

30. An element with electronic configuration 2, 8, 1 forms a compound with carbonate. How does this compound remove both temporary and permanent hardness of water?

31. In permutit process, what is the role of the addition of the concentrated solution of the solid compound obtained due to the reaction between NaOH with HCl? Explain.

32. Master Vishnu and his sister Jishnu went to their uncle's village to spend their summer vacation. One day, Jishnu was pestering her uncle for a soft drink. Vishnu's uncle took both of them to a shop where soft drink bottles were kept in an insulated box containing ice. Vishnu's uncle who is a physical science teacher, with an intention of testing Vishnu, posed a question to him regarding the concept involved in cooling of soft drinks by keeping them in ice. Vishnu who is a student of TIME, IITF impressed his uncle by explaining the above phenomenon with respect to the concept of latent heat of fusion of ice. Can you also explain the concept like Vishnu?

Level 3

1. Why is ion exchange process preferred over permutit process for the purification of water?
2. In cold countries, water pipes burst in winter. Explain.
3. Generally when a person jumps into water in any water body, he is expected to sink. But he tends to float in the Dead Sea. How do you account for this?
4. During the conversion of surface water to potable water, chlorination is an important step. Give reasons.
5. Thermal pollution of water leads to the death of biotic life. Explain.

ASSESSMENT TEST

Test 1

Directions for questions from 1 to 15

Select the correct alternative from the given choices.

1. Which of the following chemical substances can be used for removing permanent hardness of water by precipitation reaction?

   (a) baking soda (b) washing soda
   (c) alum (d) quick lime

2. Wealthy but not healthy agriculture can be achieved by

   (a) introducing biogas plants
   (b) using natural pesticides such as neem oil
   (c) using drip irrigation
   (d) using artificial fertilisers

3. Match the following:

Column A		Column B
(A) Rain water	( )	(a) Free of suspended impurities
(B) Rivers and lakes	( )	(b) Purest form of natural water
(C) Underground water	( )	(c) Water obtained after reverse osmosis
(D) Demineralised water	( )	(d) Water coming out of an ion exchange apparatus
	( )	(e) Fresh water for consumption for various activities

   (a) A → e; B → b; C → a; D → d, e
   (b) A → b; B → e; C → a; D → d
   (c) A → b; B → a; C → d; D → e
   (d) A → b; B → a, e; C → b; D → d

4. The amount of heat energy required to increase the temperature of 20 g of water by 1°C is ______.

   (a) 10 cal (b) 20 cal
   (c) 15 cal (d) 2 cal

5. Arrange the steps in a sequence for the formation of rainfall.

   (1) formation of micelles (2) condensation

(3) evaporation (4) formation of clouds
(5) precipitation
(a) 3 5 4 2 1 (b) 3 2 4 1 5
(c) 3 2 1 4 5 (d) 1 2 3 5 4

6. A divalent metal salt X, which contributes to hardness of water, combines with washing soda and forms an insoluble salt Y and common salt. Salt Y is also used for the laboratory preparation of CO_2. Identify X and Y, respectively.
(a) $CaSO_4$, $CaCO_3$
(b) $MgSO_4$, $MgCO_3$
(c) $CaCl_2$, $CaCO_3$
(d) $MgCl_2$, $MgCO_3$

7. Arrange the general steps for the purification of water in a sequence.
(1) filtration
(2) chlorination
(3) sedimentation
(4) addition of chemicals
(a) 1 3 2 4 (b) 3 1 2 4
(c) 3 4 1 2 (d) 1 2 4 3

8. Two persons A and B got burns: one with boiling water at 100°C and other due to steam at 100°C, respectively. Which person is affected more?
(a) A
(b) B
(c) both A and B to the same extent
(d) none of the above

9. Which of the following dissolved gases are present in drinking water?
(a) oxygen and nitric oxide
(b) oxygen and carbon monoxide
(c) oxygen and carbon dioxide
(d) oxygen and nitrogen

10. On passing electricity through acidulated water, the gaseous products obtained are collected in two separate test tubes A and B. The volume of the gas collected in test tube A is double the volume of gas collected in test tube B. Identify the two gases in test tubes A and B, respectively.
(a) hydrogen and oxygen
(b) oxygen and hydrogen
(c) hydrogen peroxide and oxygen
(d) oxygen and water vapour

11. Which of the following is the application of the high specific heat capacity of water?
(a) cooling of soft drink bottles in ice
(b) cooling of the earth's atmosphere after a hail storm
(c) Usage of water as a coolant
(d) bursting of water pipes during winter in cold countries

12. **Assertion (A)**: Boiling removes both temporary and permanent hardness from water.
Reason (R): Bicarbonates of calcium and magnesium undergo decomposition on heating.
(a) Both (A) and (R) are true and (R) is the correct reason for (A).
(b) Both (A) and (R) are true but (R) is not the correct reason for (A).
(c) (A) is true but (R) is false.
(d) (A) is false but (R) is true.

13. The pressure cooker reduces the cooking time because of
(a) increase in the boiling point of water
(b) high specific heat of water
(c) decrease in the boiling point of water
(d) high latent heat of vaporisation

14. **Assertion (A)**: Electrolysis of water results in liberation of hydrogen at anode and oxygen at cathode.
Reason (R): When burning splinter is introduced in a test tube at cathode, it is put off with a pop sound.
(a) Both (A) and (R) are true and (R)is the correct reason of (A).
(b) Both (A) and (R) are true but (R) is not the correct reason of (A).
(c) (A) is true but (R) is false.
(d) (A) is false but (R) is true.

15. When water gets converted into ice, its volume
(a) decreases
(b) increases
(c) remains the same
(d) initially decreases and then increases

Test 2

Directions for questions from 1 to 15

Select the correct alternative from the given choices.

1. Which of the following salts is responsible for temporary and permanent hardness of water, respectively?
   (a) $Mg(HCO_3)_2$ and Na_2SO_4
   (b) $Ca(HCO_3)_2$ and $MgCl_2$
   (c) $MgSO_4$ and $Mg(HCO_3)_2$
   (d) $CaCl_2$ and $CaSO_4$

2. Which of the following is a biodegradable substance?
   (a) detergent
   (b) pesticide
   (c) polythene
   (d) waste obtained from a biogas plant

3. Match the following:

Column A	Column B
(A) Distilled water	( ) (a) Drinking purpose
(B) Potable water	( ) (b) Laboratory work
(C) Spring water	( ) (c) Dissolved salts and rare minerals
(D) Acidulated water	( ) (d) Medicinal purposes
	( ) (e) Electrolysis

   (a) A → b; B → a; C → d, c; D → e
   (b) A → c; B → b; C → c, e; D → d
   (c) A → b; B → c; C → d; D → e
   (d) A → a; B → c; C → d, e; D → b

4. The amount of heat energy required to increase the temperature of X g of water by 10°C is found to be 15 cal. Calculate X.
   (a) 0.5 g (b) 15 g
   (c) 1.5 g (d) 7.5

5. Arrange the steps in a sequence for the conversion of atmospheric water vapour into underground water.
   (1) infiltration in recharge area
   (2) water table
   (3) infiltration in zone of aeration
   (4) precipitation
   (5) Aquifer
   (a) 4 1 3 2 5 (b) 4 3 1 2 5
   (c) 4 3 1 5 2 (d) 4 1 3 5 2

6. Water-containing salt X, of a divalent metal, when treated with a compound Y gives much lather with soap. Further, an insoluble compound $MgCO_3$ and salt of a monovalent metal sulphate are formed. Identify X and Y, respectively.
   (a) $CaCl_2$, Na_2CO_3 (b) $MgSO_4$, Na_2CO_3
   (c) $MgCl_2$, $NaHCO_3$ (d) $CaSO_4$, $Mg(HCO_3)_2$

7. Arrange the steps in a sequence for the removal of microorganisms during the purification of drinking water.
   (1) lime feeder
   (2) mechanical mixing
   (3) alum feeder
   (4) sedimentation tank
   (5) passing through gravel and sand
   (6) chlorine feeder
   (a) 3 1 2 4 5 6 (b) 3 2 1 4 5 6
   (c) 5 1 2 3 6 4 (d) 1 2 4 3 5 6

8. What is the latent heat of fusion and latent heat of vaporisation of 10 g of ice and water, respectively?
   (a) 800 cal, 800 cal
   (b) 540 cal, 540 cal
   (c) 800 cal, 5400 cal
   (d) 540 cal, 800 cal

9. Which of the following water does not contain dissolved gases such as oxygen and carbon dioxide?
   (a) potable water (b) distilled water
   (c) saline water (d) soft water

10. The volumetric composition of water is
    (a) 1 : 2 ratio of hydrogen and oxygen, respectively
    (b) 1 : 8 ratio of hydrogen and oxygen, respectively
    (c) 1 : 1 ratio of hydrogen and oxygen, respectively
    (d) 2 : 1 ratio of hydrogen and oxygen, respectively

11. The solvent water is used in car radiators. Which of the following properties of water is exploited?
    (a) high solubility
    (b) poor conductivity
    (c) maximum density
    (d) high specific heat

12. **Assertion (A)**: The Clark's method involves removal of permanent hardness.
    Reason (R): The Clark's method involves the addition of slaked lime which results in the removal of bicarbonates as insoluble carbonates.

(a) Both (A) and (R) are true and (R) is the correct reason for (A).

(b) Both (A) and (R) are true but (R) is not the correct reason for (A).

(c) (A) is true but (R) is false.

(d) (A) is false but (R) is true.

13. Which among the following is based on the principle that boiling point decreases with decrease in atmospheric pressure?

(a) Cooking food at sea level takes longer period of time.

(b) Cooking food at higher altitudes takes longer period of time.

(c) Working principle of air coolers.

(d) Working principle of a pressure cooker.

14. **Assertion (A)**: The products of electrolysis of water are hydrogen and oxygen.

Reason (R): In electrolysis process of water, hydrogen ions are attracted to anode and hydroxyl ions are attracted to cathode.

(a) Both (A) and (R) are true and (R) is the correct reason of (A).

(b) Both (A) and (R) are true but (R) is not the correct reason of (A).

(c) (A) is true but (R) is false.

(d) (A) is false but (R) is true.

15. When the temperature of water rises from 0°C to 10°C, the density of water

(a) decreases gradually

(b) decreases up to 4°C and then increases

(c) increases up to 4°C and then decreases

(d) increases gradually

TEST YOUR CONCEPTS

Fill in the blanks

1. 97 per cent
2. sodium, magnesium, calcium
3. carbonic acid
4. 540
5. 1 cm^3
6. scum
7. two, one
8. Temporary
9. Cetyl alcohol
10. drip irrigation

Multiple choice questions

11. d
12. a
13. b
14. a
15. a
16. d
17. b
18. b
19. d
20. c
21. b
22. c
23. b
24. c
25. b

Match the following

26. A → c, d
    B → e
    C → a
    D → b

27. A → e
    B → d, c
    C → a
    D → b

28. A → b
    B → c
    C → d
    D → a

CONCEPT APPLICATION

Level 1

True or false

1. True
2. True
3. False
4. True
5. False
6. True
7. True
8. False
9. False
10. True

Fill in the blanks

11. Well water, spring water
12. permutit
13. Ca^{+2}, Mg^{+2}
14. hastens
15. bleaching powder, chlorine
16. increases
17. hydrogen
18. evaporation
19. micelles
20. 70 per cent, 90 per cent

Multiple choice questions

21. d
22. b
23. b
24. a
25. b
26. d
27. d
28. c
29. d
30. a
31. d
32. b
33. b
34. c
35. a
36. a
37. a

Match the following

38. A → c
    B → a
    C → d
    D → b

39. A → c
    B → d
    C → b
    D → a

40. A → d
    B → e
    C → f
    D → b
    E → c

Level 2

Multiple choice questions

1. c	2. c	3. d	4. d	5. c
6. b	7. a	8. d	9. a	10. d
11. b	12. d	13. c	14. 3	15. b
16. c	17. a	18. b		

ASSESSMENT TEST

Test 1

1. b	2. d	3. b	4. b	5. c
6. c	7. c	8. b	9. c	10. a
11. c	12. d	13. a	14. d	15. b

Test 2

1. b	2. d	3. a	4. c	5. b
6. b	7. a	8. c	9. b	10. d
11. d	12. d	13. b	14. c	15. c

CONCEPT APPLICATION

Level 1

True or false

1. Under ground water is passed through several layers of soil, and hence, it is free from suspended impurities.
2. The water which contains large amounts of dissolved salt is called saline water.
3. 1 g of pure water at 4°C occupies a volume of 1 cc.
4. Soft water produces lather with soap, whereas hard water produces scum with soap.
5. Hard water is unfit for dyeing clothes as it makes colour to fade.
6. Cation exchanger is regenerated by passing HCl solution.
7. Pure water is a bad conductor of electricity. Salt or alkali is added to water to increase its conductivity.
8. $2H_2 + O_2 \rightarrow 2H_2O$

   Two volumes, one volume, two volumes

9. Clark's method is used for the removal of temporary hardness.
10. Specific heat capacity of water is 1 cal/g/°C.

Fill in the blanks

11. Well water and spring water are the two types of underground water.
12. Hydrated sodium aluminium silicate is called permutit.
13. The exhausted permutit contains Ca^{+2} and Mg^{+2} in place of Na^{+} ions.
14. In sedimentation process, alum hastens the process of precipitation of suspended impurities.
15. Addition of bleaching powder releases chlorine which helps in killing the harmful bacteria present in water.
16. Addition of acid to water increases the conductivity.
17. In electrolysis of water, hydrogen gas is liberated at cathode.
18. The change of water into water vapour below 100°C at slow rate is called evaporation.
19. Small water droplets associated with the dust particles form larger aggregates called micelles.
20. Percentage of water in animals varies from 70 to 90 per cent.

Multiple choice questions

21. Distillation process is not involved in the purification of drinking water.
22. Sea water is not used for drinking purpose due to the presence of dissolved salts.
23. Density of water is maximum at 4°C.
24. Adding of manure to soil causes less pollution because it is an organic matter and can be decomposed by microorganisms.
25. Metals present in permutit are Na and Al.
26. Water has the highest specific heat among all the substances.
27. Distilled water is a poor conductor of electricity as it is the purest form of water.
28. Usage of fertilisers is not the control measure for water pollution.
29. Nitrate and phosphate ions do not cause hardness of water.
30. Anion exchange resin can be regenerated by passing sodium hydroxide solution because it can replenishes OH^- ions.

31. Spring water is used for medicinal purposes because it contains dissolved salts, and minerals but not suspended impurities.

32. Potable water, soft water and sea water contain dissolved gases such as oxygen and carbon dioxide but distilled water does not contain O_2 and CO_2 because during distillation these dissolved gases are distilled off.

33. Washing soda is used to remove permanent hardness by precipitation reaction.

34. Density of water is maximum at 4°C due to anomalous expansion of water. Hence, when the temperature of water is raised from 0°C to 10°C, the density increases up to 4°C, then decreases gradually.

35. The volumetric composition of water is 2 : 1 of hydrogen and oxygen.

36. (i) lime feeder, alum feeder

(ii) mechanical mixing

(iii) filtration

(iv) chlorination

37. (i) ice of 10 g mass at 10°C

(ii) ice of 10 g mass at 40°C

(iii) water of 10 g mass at 10°C

(iv) water of 10 g mass at 40°C

(v) steam of 10 g mass at 10°C

Match the following

38. A → c Defecation by living beings produces pollutants such as amoeba, fungicides, etc.

B → a Industrial operation produces pollutants such as mercury, arsenic, etc.

C → d Water released from thermal power plants is favourable for growth of algae.

D → b Agricultural activities produce pollutants such as insecticides, fungicides, etc.

39. A → c By the process of boiling we can remove temporary hardness in which bicarbonates present in water decompose to carbonates which are precipitated.

B → d Clark's method involves the addition of slaked lime to water either in solid or liquid form to remove temporary hardness.

C → b Addition of washing soda helps in removing both temporary and permanent hardness.

D → a Ion exchange method consists of solids called ion exchange resins.

40. A → d The value of specific heat of water is 1 cal/g/°C.

B → e The latent heat of fusion of ice is 80 cal/g/°C.

C → f The latent heat of vaporisation of water is 540 cal/g/°C.

D → b Ice melts at 0°C.

E → c Water boils at 100°C.

Level 2

1. Latent heat of fusion of 10 g of water is 10 × 80 = 800 cal

Latent heat of vaporisation of 10 g of steam is 10 × 540 = 5400 cal

2. Icebergs weigh millions of tons but float on the surface of sea water because the density of ice is less than that of the sea water. Usually 9/10 of icebergs will be below the surface of water and 1/10 above the sea water as a result we cannot estimate the size below water, and hence, when ship hits this heavier ice berg the ship sinks.

3. Distillation process is not involved in making surface water potable.

4. Water has high specific heat, which means that water needs a large amount of heat to become hot.

5. Divalent metal salt X which contributes to hardness of water is $CaCl_2$ which combines with Na_2CO_3 gives insoluble salt Y, i.e., calcium carbonate and sodium chloride.

$$CaCl_2 + Na_2CO_3 \rightarrow CaCO_3\downarrow + 2NaCl$$

6. Specific heat of water is 1 cal/g/°C.

That means 1 g of water requires 1 cal energy to increase the temperature by 7°C.

2 g of water requires? = 20 × 1 = 20 cal

7. On passing electricity through acidulated water it decomposes to hydrogen and oxygen. Volume of hydrogen evolved in test tube A is double the volume of oxygen collected in test tube B.

8. Demineralised water is obtained through ion-exchange process. Distilled and demineralised water doesn't contain any impurities. Hence, they do not conduct electricity. Mineral water is a good conductor of electricity.

9. Cooking in a pressure cooker takes less time because in pressure cooker as pressure increases boiling point increases and food is cooked softer in less time.

10. By introducing artificial fertilisers, crop yield is improved but at the same time it contributes to water pollution.

11. B will get severe burns as latent heat of vaporisation of 540 cal/g is stored within the molecules of steam.

12. Aquatic organisms are able to survive in lakes during winter even though there is ice at the surface because the density of ice is less than that of water; ice is a poor conductor of heat.

13. Chlorine is involved in making surface water potable by the process called chlorination.

14. In air coolers, warm air is sucked by an exhaust fan. The warm air passes through khus khus mats through which water is dripping. The water absorbs the heat from the air and cools the air.

15. Divalent metal salt X, i.e., $MgSO_4$ is treated with compound Y, i.e., Na_2CO_3 to form $MgCO_3$ and Na_2SO_4.

$$MgSO_4 + Na_2CO_3 \rightarrow MgCO_3 + Na_2SO_4$$

16. The specific heat of water is 1 cal/g/°C. Therefore, the amount of heat required is 1.5 cal for 15 for 10°C raise in temperature.

17. In the ion-exchange method, pure water called demineralised water is obtained which is not potable because it does not contain any minerals and dissolved salts.

18. As we move to higher altitudes, pressure decreases, and hence, boiling point decreases. Water boils at low temperatures on the hills, and thus, the food is cooked for longer hours.

19. With the increase in altitude the atmospheric pressure decreases. Boiling point of water is directly proportional to the atmospheric pressure. Hence, boiling point of water decreases at the top of a hill and cooking takes a longer period of time.

20. Since the density of solid wax is greater than the molten wax, molten wax will overflow. Density of ice is less than that of water. As a result, water formed from ice occupies lesser volume instead of overflowing.

21. Sea water contains more amount of dissolved salts, and hence, it is unsuitable for drinking. Sea water is made potable through reverse osmosis process.

22. The specific heat capacity of water is very high. Hence, it can lose lot of heat without much decrease in its temperature. The heat lost by water can increase the temperature of the other substances which is in contact with it. Therefore, hot water bags can used to keep out body warm in winter.

23. Density of ice is less than water, and as a result ice floats on the surface of water.

24. Specific heat capacity of water is 1 cal/g/°C.

1 cal heat is required to increase the temperature of 1 g of water by 1°C.

To increase the temperature of 10 g water by 12°C, 120 calories of heat is required.

25. Specific heat capacity of water is very high. Hence, water can take way lot of heat from the hot engine without significant increase in its temperature.

26. Latent heat of fusion or freezing is 80 cal/g, i.e., 1 g of ice requires 80 cal.

20 g ice requires? = 20 + 80 = 1600 cal

Latent heat of vaporisation of water is 540 kg cal/g, i.e., 1 g water requires 540 cal.

20 g water requires? = 20 × 540 = 10,800 cal

Ratio = 1600 : 10800 = 1 : 6.75

27. Addition of cetyl alcohol reduces the rate of evaporation of water. Hence, level of water in container A is comparatively higher than the level of water in container B.

28. Elements X and Y are Mg and Ca, respectively. Hydrogen carbonates or bicarbonates of Ca and Mg impart temporary hardness while sulphates of Ca and Mg impart permanent hardness to water.

29. The heat energy supplied is completely utilised to convert ice into liquid water. As a result, thermometer does not register any rise in temperature till all the ice cubes melts.

30. Element with electronic configuration 2, 8, 1 is Na and its carbonate is sodium carbonate which is called washing soda. It is used for the removal of both temporary and permanent hardness of water.

$Ca(HCO_3)_2 + Na_2CO_3 \rightarrow CaCO_3 \downarrow + 2NaHCO_3$

$CaSO_4 + Na_2CO_3 \rightarrow CaCO_3 \downarrow + Na_2SO_4$

$CaCO_3$ is insoluble, and hence, it settles down.

In this way washing soda can be used to remove both temporary and permanent hardness of water.

31. In permutit process, we use sodium permutit, i.e., $Na_2Al_2Si_2O_g.xH_2O$. When hard water is treated with sodium premutit, Ca and Mg ions are exchanged with the Na ions present in the permutit. After long use, most of the sodium ions present in permutit are replaced with calcium and magnesium ions, and hence, it becomes ineffective. Sodium hydroxide reacts with HCl to from sodium chloride Concentrated sodium chloride solution is added to regenerate the permutit.

32. The amount of heat energy required to convert 1 g of ice to 1 g of water at 0°C is called latent heat of fusion. Latent heat of fusion of ice is 80 cal/g, i.e., every 1 g of ice requires 80 calories of heat in order to melt. This energy is taken away from the soft drinks, and hence, these drinks are cooled when they are placed in ice.

Level 3

1. In permutit process, cations which impart hardness to water are removed. In the ion-exchange process both cations and anions are removed in the respective ion exchanger. Hence, water obtained from the ion exchanger is almost free from ions which can be used in the laboratory to carry out experiments.

2. In cold countries, the temperature is below 0°C. The water present in the pipes expands during freezing. As a result, ice exerts huge pressure on the pipes and they burst.

3. Dead sea consists of large quantity of dissolved salts to the extent of saturation. Density of human body is less than the water present in dead sea. Therefore, human body floats.

4. During chlorination, chlorine is added to water, which produces nascent oxygen. This nascent oxygen kills all harmful germs and bacteria present in water.

5. Thermal pollution of water is due to heavy industries such as steel, thermal power plants. These sources discharge large amounts of hot water into the rivers, which helps in the rapid growth of algae. This alga depletes dissolved oxygen in water. This in turn leads to the death of biotic life.

ASSESSMENT TEST

Test 1

1. Washing soda can be used for removing permanent hardness by precipitation reaction.

2. By introducing artificial fertilisers, crop yield is improved but at the same time it contributes to water pollution.

3. A → b Rain water is the purest form of natural water.

   B → e River and lake water are considered as fresh water, and hence, available for consumption for various activities.

   C → a Underground water during seepage gets filtered through various layers of soil, and hence, it does not contain suspended impurities.

   D → d The water coming out of an ion-exchange apparatus contains only pure water and is called dematerialised water.

4. Specific heat of water is 1 cal/g °C.

   Amount of heat energy required is 20 × 1 = 20 cal.

   1 cal – 1 g

   ? – 20 g = 20 cal

5. (i) Evaporation

   (ii) Condensation

   (iii) Formation of micelles

   (iv) Formation of clouds

   (v) Precipitation

6. Divalent metal salt X which contributes to hardness of water is $CaCl_2$ which combines with Na_2CO_3. It gives salt Y, calcium carbonate and sodium chloride.

   $CaCl_2 + Na_2CO_3 — CaCO_3 + 2NaCl$

7. (i) Sedimentation

   (ii) Addition of chemicals

   (iii) Filtration

   (iv) Chlorination

8. B will get severe burns as latent heat of vaporisation of 540 cal/g is stored within the molecules of steam.

9. Oxygen and carbon dioxide are the two dissolved gases present in drinking water.

10. On passing electricity through acidulated water it decomposes to hydrogen and oxygen. Volume of hydrogen evolved in test tube A is double the volume of oxygen collected in test tube B.

11. Water can be used as a coolant because of its high specific heat.

12. Boiling removes only temporary hardness because soluble bicarbonates of calcium and magnesium undergo decomposition on heating and insoluble carbonates are formed.

13. Cooking in a pressure cooker takes less time because in the pressure cooker as pressure increases boiling point increases.

14. In the electrolysis of water, hydrogen is liberated at cathode and oxygen is liberated at anode. When a burning splinter is introduced into the test tube with hydrogen, it is put off with a pop sound.

15. When the water gets converted into ice, its volume increases as density of ice is less than that of water.

Test 2

1. Bicarbonates of calcium and magnesium and chlorides and sulphates of calcium and magnesium are responsible for temporary and permanent hardness, respectively.

2. The waste obtained from biogas plant can be used as manure as it is an organic matter.

3. A → b Distilled water is used in a laboratory for chemical reactions taking place in water.

HINTS AND EXPLANATION

B → a Potable water is used for drinking purpose.

C → d, c Spring water is devoid of suspended impurities but contains dissolved salts and minerals.

Due to the presence of rare minerals, it is used for medicinal purposes.

D → e Pure water is a bad conductor of electricity, little amount of sulphuric acid is added to make acidulated water.

4. The specific heat of water is 1 cal/g/°C.

∴ Heat required for increasing 10°C temperature of X g water is 10 X cal.

∴ 10X = 15

∴ X = 1.5 g

5. (i) Precipitation

(ii) infiltration in zone of aeration

(iii) infiltration in recharge area

(iv) water table

(v) aquifer

6. Divalent metal salt X, i.e., $MgSO_4$ is treated with compound Y, i.e., Na_2CO_3 to form $MgCO_3$ and Na_2SO_4.

$$MgSO_4 + Na_2CO_3 \rightarrow MgCO_3 + Na_2SO_4$$

7. (i) alum feeder

(ii) lime feeder

(iii) mechanical mixing

(iv) sedimentation tank

(v) passing through gravel and sand

(vi) chlorine feeder

8. Latent heat of fusion of 10 g of ice is 10 × 80 = 800 cal

Latent heat of vaporisation of 10 g of water is 10 × 540 = 5400 cal.

9. Potable water, soft or sea water contain dissolved gases such as oxygen and carbon dioxide but distilled water does not contain O_2 and CO_2 because during distillation these dissolved gases are distilled off.

10. The volumetric composition of water is 2 : 1 of hydrogen and oxygen.

11. Water has high specific heat, which means that water needs a large amount of heat to become hot.

12. Clark's method is used to remove temporary hardness. In this process slaked lime is added, which results in conversion of soluble bicarbonates into insoluble carbonates. On filtration, soft water is obtained.

13. As we move to higher altitudes, pressure decreases, and hence, boiling point decreases. Water boils at low temperatures on the hills, and thus, the cooking of food takes longer time.

14. During the electrolysis of water the electrolytes in water dissociates as H^+ and OH^- ions which move to cathode and anode, respectively. Oxygen and hydrogen are produced.

15. Density of water is maximum at 4°C due to anomalous expansion of water. Hence, when the temperature of water is raised from 0°C to 10°C, the density increases up to 4°C, then decreases gradually.

Chapter 6

Chemistry in Daily Life

Remember

Before beginning this chapter, you should be able to:

- understand basic concepts of matter and its classification.
- remember physical and chemical properties of certain substances.

Key Ideas

After completing this chapter, you should be able to:

- understand the properties of metals and non-metals, and their applications in various fields.
- find out some common alloys, their special properties and uses.
- learn about the preparation, properties of different compounds, like, plaster of Paris, cement, glass, fertilisers, soaps and detergents that find wide applications in our daily life.
- understand different types of polymers and plastics.

INTRODUCTION

Chemistry is the branch of science which deals with the composition, physical and chemical behaviour of different substances around us. It also deals with the application of these substances in various fields to make our lifestyle easy and comfortable. Apart from this, numerous chemicals are synthesized in the laboratory and then in the industry to satisfy different requirements. Moreover, improvisation of the properties of the substances available in nature by altering their composition also comes under the purview of chemistry. Hence, the study of chemistry is indispensable. A brief overview of different chemical substances, like, metals, non-metals, alloys, compounds of different elements, mixtures, their properties and the area of applications is discussed in this chapter.

Metals and their Uses

Metals find extensive applications because of their specific properties, like-lustre, tensile strength, conductivity, malleability, ductility, etc.

TABLE 6.1 Properties and applications of different metals

Name of the metals	Properties	Applications
Gold	1. It possesses a yellow lustre, is malleable, ductile and highly resistant to chemicals. It is relatively a soft metal.	1. It is used for making jewellery, coins, etc. 2. Used in dentistry for filling cavities, satellites and ayurvedic medicines.
Silver	1. It possesses a white lustre, is malleable, ductile and is one of the best conductors of heat and electricity.	1. It is used for making coins and ornaments, table cutlery, high quality glass mirror, silver threads in clothing and electric water purifiers. 2. It is used in electric circuits for highly expensive electric appliances.
	2. Compounds of silver, like, silver chloride, silver iodide, etc., are sensitive to light and undergo photolytic decomposition.	It finds application in ayurvedic medicine and filling cavities in teeth. Compounds of silver, such as, silver chloride and silver iodide are used in photographic film and photogrey lenses. Silver nitrite or silver iodide is used in cloud seeding for artificial rain.
Copper	1. It possesses a reddish brown lustre, is malleable, ductile and is the second best conductor of heat and electricity.	1. It is used for the manufacture of electric transmission cables, utensils, statues, etc. 2. It is used for making printed circuits used in electronic devices. 3. It is also used for making coins and alloys like, brass, bronze, bell metal, etc.
Iron	1. It possesses a grey lustre and is fairly malleable and ductile.	1. It is extensively used in the form of an alloy in the construction of ships, buildings, automobiles, railway bridges, railway lines, electric transmission towers, television towers, microwave towers and in a variety of machineries. 2. It finds wide applications in household articles.
Tin	1. It possesses a silvery white lustre, is malleable and less ductile. It has a comparatively low melting point with respect to other metals.	1. It is used for coating food cans and cooking vessels which are made of copper and iron. 2. The process of coating tin over any article is called **tinning**. 3. It is used in the manufacture of alloys like brass & bronze and fusible alloys which are used in the manufacture of fuse wires.

Name of the metals	Properties	Applications
Lead	1. It possesses a silvery grey lustre, is less corrosive, and hence, more durable. It has a comparatively lower melting point and its density is very high.	1. It is used for making lead storage batteries used in automobiles. 2. It is used for covering tin roofs in cold countries, for the manufacture of sanitary pipes and making the tip of bullets. 3. It is used for making fusible alloy constituted of lead and tin in 1 : 1 ratio by mass.
Zinc	1. It possesses a white lustre with a light blue tinge. 2. Characteristic properties of metals are, namely, malleability and ductility are not clearly demonstrated in zinc.	1. It is used in the process of galvanising iron sheets. In the galvanisation process, the hot iron sheets are dipped in molten zinc and then passed through hot and heavy rollers. On cooling, a coating of zinc metal is formed over the iron sheets. The coating of zinc protects iron from rusting, and hence, is used for the manufacture of iron roofing, iron tanks, rainwater drain pipes, etc. 2. It is used as electrodes in dry cells. 3. It is used for making alloys, such as, brass and bronze.
Aluminium	1. It possesses a silvery white lustre, is highly malleable and ductile, is a very good conductor of electricity, is very light in weight and corrosion resistant. 2. It is the most abundant metal in the earth's crust. 3. It is relatively a soft metal when compared to other metals.	1. It is used for the manufacture of high voltage electric transmission wire. 2. It is used for making foils for packing food materials, medicines, etc. 3. It is used for making alloys, like, duralumin, magnalium, etc., which are very light in weight and impact resistant. Hence, these alloys are used for the manufacture of aircraft frames and components of automobiles. 4. These alloys also find application in the manufacture of utensils, household fixtures, etc.
Magnesium	1. It possesses a silvery white lustre, burns with a white dazzling flame and its reaction with oxygen is highly exothermic.	1. It is used in fire works. 2. It is used as fuse wires in thermite welding. 3. It is used in the manufacture of alloys, like, magnalium.
Mercury	1. It is a silvery white metal which is a liquid at normal temperature. Its melting point is –39°C and boiling point is 357°C, is a very good conductor of electricity and heat. 2. It can dissolve almost all the metals in itself. The solution of a metal in mercury is called amalgam.	1. It is used as a thermometric liquid in the laboratory as well as a clinical thermometer. 2. The gold amalgam and silver amalgam are used in dentistry for filling teeth cavities.

TABLE 6.2 Properties and applications of different non-metals

Names of the non-metals	Properties	Applications
Chlorine	1. It is a greenish yellow, pungent smelling and poisonous gas. 2. It is fairly soluble in water and about 2.5-times heavier than water. 3. It shows bleaching (decolourizing) and sterilising action.	1. It is used for bleaching cotton and jute fabrics. 2. It is used for the sterilisation of drinking water. 3. It is also used in the manufacture of bleaching powder, hydrochloric acid, insecticides and pesticides, like, DDT (dichloro diphenyl trichloroethane) and BHC (benzene hexachloride).

Names of the non-metals	Properties	Applications
Iodine	1. It is a bluish-grey crystalline solid with lustre. 2. It is insoluble in water, but soluble in alcohol, carbon tetrachloride and potassium iodide solution.	1. Compounds of iodine, like, sodium iodide or potassium iodide are mixed with common salt to prevent the deficiency of iodine in our diet. 2. The common salt containing the above compounds of iodine is called iodised salt. People suffer from diseases, like, goitre and mental retardation because of the deficiency of iodine in their body. 3. Other compounds of iodine, like, silver iodide are used for making photographic films as well as for cloud seeding. 4. Tincture of iodine (iodine in mixture of potassium iodide solution and alcohol) is used as a germicide for dressing of wounds. 5. Iodoform, another compound of iodine, is used in medicines.
Graphite	1. It is a crystalline form of carbon. It possesses a dark grey lustre. 2. It is a good conductor of electricity and bad conductor of heat. 3. It is highly heat resistant as well as chemical resistant. 4. It is soft, and hence, can leave an impression on the paper.	1. Pencil leads are manufactured by mixing graphite with clay. The hardness of the pencil depends on the amount of clay present in it. The more the amount of clay, the harder is the pencil. 2. Graphite mixed with fine clay is moulded and sintered into crucibles of desired size and shape. These crucibles are used to melt small quantities of metals. 3. It is also used as an excellent lubricant in those parts of the machineries where high heat is generated. In the above case, the use of ordinary lubricating oils is restricted as they get charred at high temperatures. 4. Graphite is used in electrolytic cells as electrodes. 5. Graphite is used in nuclear reactors as a moderator to slow down the speed of neutrons produced due to the splitting of the atoms of uranium or thorium.

Noble Gases and their Uses

Noble gases find various applications due to their **inert (non-reactive)** nature. There are six noble gases which are **helium, neon, argon, krypton, xenon** and **radon**. Helium, being the second lightest element and inert in nature, is extensively used for filling weather observation balloons. Neon is used in advertising sign boards, whereas argon is used in filling up the electric bulbs to improve their efficiency by preventing oxidation of tungsten filament. Krypton and xenon are also used in electric bulbs to slow down the sublimation of hot tungsten filament. Radon, being a radioactive gas is used in cancer treatment.

Some Common Alloys and their Uses

In order to improvise the properties of a metal according to the end use, fused or molten metal is uniformly mixed with other metal(s) and sometimes a non-metal in small quantity. The mixture is then cooled to obtain a solid, called an alloy. Every alloy possesses some special properties because of its specific composition. In general, alloying improves the mechanical strength, makes the metal corrosion resistant, and hence, increases the durability.

TABLE 6.3 Important alloys and their uses

Names of the alloys	Major components	Composition (other components)	Special properties	Applications
Stainless steel	Iron (83%)	Carbon (1%) Chromium (15%) Nickel (1%)	It does not rust and is resistant to mild acids or alkalies.	In utensils, cutlery, surgical instruments and ornamental articles.
Duralumin	Aluminium (95%)	Copper (4%) Manganese (0.5%) Magnesium (0.5%)	It is very light in weight but hard and possesses high tensile strength.	It is used for making frames of aircraft, automobiles and speedboats.
Brass	Copper (60–80%)	Zinc (40–20%)	It possesses a yellow lustre, is harder and stronger than copper but more malleable and ductile than copper.	It is used for making electric switches of various kinds, taps, statues, household utensils and shells of ammunition rounds.
Bronze	Copper (80%)	Zinc (10%), Tin (10%)	It is resistant to corrosion and takes high polish.	It is used for making statues, coins and household utensils.
Alloyed gold	Gold	Silver or copper in small quantities	It possesses a lower melting point and is harder and more impact resistant than pure gold.	It is extensively used in making jewellery and coins.

EXAMPLE

What are noble gases? Mention their uses.

SOLUTION

Noble gases are those which are inert. There are six noble gases which are helium, neon, argon, krypon, xenon and radon.

	Element	Uses
1.	Helium	Filling weather observation balloons
2.	Neon	Advertising sign boards
3.	Argon, krypton and xenon	Filling up electric bulbs
4.	Radon	Cancer treatment

EXAMPLE

Describe the process of galvanisation. How is it useful?

SOLUTION

The process of coating zinc over iron articles is called galvanisation. Zinc is used in the process of galvanising iron sheets. In the galvanisation process, the hot iron sheets are dipped in molten zinc and then passed through hot and heavy rollers. On cooling, a coating of zinc metal is formed over the iron sheets. The coating of zinc protects iron from rusting, and hence, is used for the manufacture of iron tanks, rain water, drain pipes, etc.

EXAMPLE

(a) What is the composition of pencil lead?

(b) How is hardness of pencil related to the composition of pencil lead?

(c) How are crucibles made with the help of graphite?

(d) Why is graphite a more preferred lubricant to the ordinary lubricating oils?

(e) What is the role of graphite in nuclear reactor?

SOLUTION

(a) Pencil leads are manufactured by mixing graphite with clay.

(b) The more the amount of clay, the harder is the pencil.

(c) Graphite mixed with fine clay is moulded and sintered into crucibles of desired size and shape.

(d) Graphite is highly resistant to heat, and therefore, used as lubricant in those parts of machinery where high heat is generated due to friction. In the above case, ordinary lubricating oils get charred at high temperatures.

(e) Graphite is used in nuclear reactors as a moderator to slow down the speed of neutrons produced due to splitting of the atoms of uranium or thorium.

EXAMPLE

An element X is non-malleable and non-ductile. X is extensively used to protect another element Y from corrosion. Identify X and Y and name the process that is mentioned above. Also compare the reactivity of X and Y.

SOLUTION

Element X is zinc which is non-malleable and non-ductile. Y is iron. The process of protecting iron by coating with zinc is called galvanisation. The element X(Zn) is more reactive than Y (Fe).

EXAMPLE

Explain the principle involved in the usage of chlorine in sterilisation of drinking water.

SOLUTION

Chlorine is highly soluble in water. On dissolution of chlorine in water, hypochlorous acid is formed which dissociates to give nascent oxygen. Nascent oxygen kills germs and bacteria. Hence, chlorine is used in sterilisation of drinking water.

Some Useful Compounds

In compounds, the constituent elements do not retain their properties and a new set of properties are exhibited by the compounds. Many compounds find applications in various fields. We now discuss some of these compounds.

Sodium chloride: The major source of sodium chloride is sea water. The salt obtained due to the evaporation of sea water is called crude salt, because of the presence of sand and clay particles. Crude salt is dissolved in water, and thus, the soluble component, i.e., sodium chloride is extracted and then recrystallised. **Recrystallisation** is a process to separate a solid solute from the liquid solvent by cooling the hot saturated solution.

A small quantity of sodium iodide or potassium iodide is mixed with sodium chloride and the mixture is called common salt, which is inevitable in our daily meal. Other than this, it is used as a preservative for the storage of fish and meat. It is also used as a raw material for the manufacture of other useful compounds, like, baking soda, washing soda, hydrochloric acid, caustic soda and elements like chlorine.

Hydrochloric acid: Hydrochloric acid is prepared industrially by heating common salt and concentrated sulphuric acid. It can also be obtained by passing hydrogen chloride gas through water. It is used in the preparation of silver chloride which finds application in photographic printing paper. It is an important laboratory reagent. Industrially, it is used to clean the surface of metals. In households, hydrochloric acid is used as a bathroom cleaning agent.

Blue vitriol: Blue vitriol is hydrated copper sulphate prepared by dissolving cupric sulphide or cupric oxide or cupric carbonate in dil sulphuric acid. It finds application as an insecticide. Solution of copper sulphate is sprayed over citrus trees and grape vines to kill the insects.

Marble or limestone: Limestone is a metamorphic rock formed from the sea shells. Chemically, limestone is calcium carbonate. Limestone is used as one of the raw materials for the manufacture of cement and glass. Lime (calcium oxide) which is a component of mortar, is manufactured from limestone. Mortar is used in the construction of buildings. Lime is also used for white washing. Marble is extensively used as flooring, statues and monuments.

Zincite or zinc oxide: Naturally occurring zinc carbonate is heated strongly to obtain zinc oxide. It is used as a raw material for the extraction of zinc and in the manufacture of paints.

Gypsum and plaster of Paris: Gypsum is a naturally occurring mineral. Its chemical composition is **calcium sulphate dihydrate ($CaSO_4.2H_2O$)**. It is used to reduce the acidity of the soil. When gypsum is heated over a long period of time, it partially loses its water of crystallisation and **plaster of Paris ($CaSO_4.½H_2O$)** is obtained.

Plaster of Paris can be moulded into different shapes when it is kneaded with the right proportion of water. The moulded object sets into a hard mass when it is dried. Hence, it is applied on the body parts to set fractured bones properly. It is also used for casting statues, toys and decorative materials. Modern black board chalks and ceiling panels are also made from plaster of Paris.

Compounds of Silicon

Sand or silicon dioxide is a naturally occurring mineral. Sand in purified form is used as the raw material for the manufacture of glass and cement.

Sodium silicate is the compound of silicon which is soluble in water and is used in the preservation of eggs. **Silicone** is a polymer manufactured in the industry from silicon. Silicone is highly heat resistant, non-sticky in nature and water repellant as well as a bad conductor of electricity. Hence, it is used in the manufacture of non-stick pans, waterproof clothes, wax polishes, electrical insulators, etc.

Ferrosilicon is an alloy of iron and silicon and is a very hard steel. **Silicon carbide**, a compound of silicon, is the hardest compound and used in all kinds of grinding wheels.

Different Types of Fertilisers and Their Uses

Fertilisers are compounds or mixtures of compounds which are added to the soil in order to provide the essential nutrients to the plants.

Fertilisers can be categorised into two types depending on the source from which they are obtained.

1. Natural fertiliser
2. Artificial fertiliser

Natural fertiliser: An example of a natural fertiliser is manure.The excreta of live stock and humans is mixed with straw and then the mixture is buried deep into the pits for few months. The mixture decomposes there and manure is obtained.The manure is rich in the nutrients required for the growth of the plants.

Artificial fertiliser: Artificial fertilisers are the chemical compounds manufactured in the industries. They contain one or more essential nutrients required for the growth of the plant. Artificial fertilisers can further be classified depending on their chemical composition, i.e., the essential nutrient(s) they contain.

1. **Nitrogenous fertiliser:** These fertilisers replenish the deficiency of nitrogen. Calcium ammonium nitrate, ammonium sulphate, basic calcium nitrate and urea are the examples of nitrogenous fertilisers.
2. **Phosphatic fertiliser:** These fertilisers replenish the deficiency of phosphorus. Triammonium phosphate and superphosphate of lime are the examples of phosphatic fertilisers.
3. **Potash fertiliser:** These fertilisers replenish the deficiency of potassium. Potassium nitrate, potassium sulphate and potassium chloride are the examples of potash fertilisers.

Some Organic Compounds used as Food

Different biological processes that are carried out in living beings require energy. In case of animals, the source of this energy is the food they consume.The food consumed by the animals can be broadly categorised into carbohydrates, proteins and fats based on the chemical composition.

TABLE 6.4 Different food materials

Types of food	Constituent elements	Examples	Uses
Carbohydrates	Composed of carbon, hydrogen and oxygen.	Glucose ($C_6H_{12}O_6$), sugar ($C_{12}H_{22}O_{11}$), starch ($(C_6H_{10}O_5)n$), etc. All cereals contain large quantity of carbohydrates.	Carbohydrates which we consume break into simpler molecules, i.e., glucose. Glucose molecules are carried to the cells with the help of blood. During respiration, glucose decomposes into carbon dioxide and water and energy are liberated.
Fats	Mainly composed of carbon, hydrogen and oxygen.	Vegetable oil, animal fat, etc.	Fat provides the maximum amount of energy.
Protein	Mainly composed of carbon, hydrogen, oxygen and nitrogen. Sulphur and phosphorus may also be present in protein.	Salivary amylase, keratin, albumin, etc.	Protein is not only the source of energy but also plays a major role in the development of cells, growth of the new cells and repair of the damaged cells.

The organic compounds which contain only carbon and hydrogen are called **hydrocarbons**.These compounds on combustion give large amount of energy, and hence, are used as fuels. Methane is the simplest hydrocarbon with the formulae CH_4.

It is the major component of natural gas. It occurs in nature as underground deposit. The decomposition of farm waste and cow dung in the absence of air gives a mixture of gases called biogas. Methane is also the major component of biogas. It is an excellent fuel with a ratio of energy given out by the combustion of methane to its mass being very high compared to the other fuels. Methane liberates carbon dioxide and water vapour on burning. It neither liberates any harmful gas, nor any solid residue. It finds application in households as well as in automobile fuels.

Some Common Solutions in Our Daily Life

Our body needs an adequate amount of fluid to carry the digested food to the cells. In the cells, the digested food gives out energy by the process of respiration. Sometimes the quantity of fluid in our body decreases due to tedious activity or some infection. The loss of fluid can be supplemented by drinking solutions, like, soda water, soft drinks, syrups or solution of glucose. The solution of glucose can be injected directly into the blood if it is urgently required.

Soda water or soft drinks: Soda water is prepared by passing carbon dioxide through water under 30 atm pressure. Excess amount of carbon dioxide gets dissolved in water due to high pressure. If the carbon dioxide is passed through a sugar solution flavoured with desired substances, the solution obtained is called soft drink.

Syrups: Highly concentrated sugar solution flavoured with desired substances is called syrup. When syrup is diluted with water, a drink rich in energy is obtained.

Glucose saline: A person suffering from dysentery may get dehydrated, i.e., excessive loss of water from the body may take place. Then a solution of glucose is directly injected into the blood to provide the required water and energy, immediately. The solution of glucose can be taken orally to get instant energy because glucose is the simplest carbohydrate.

Some Important Industrially Useful Materials

Certain mixtures, like, lime, mortar, cement, glass, etc., are manufactured in the industries through a series of processes. These substances not only fulfill our essential requirements but also they find applications even in critical scientific and biomedical researches.

Lime Mortar

It is a construction material prepared by mixing calcined lime and sand. This mixture is converted into thick slurry by adding required quantity of water. It is used for joining bricks and plastering.

Calcined Lime

When lime ($CaCO_3$) is heated to about 1000°C, CO_2 is liberated leaving behind dry lime. This dry lime is called calcined lime.

Cement

Cement is a powdery substance and greenish grey in colour. It forms a hard mass when it is mixed with water, and dried. The major area of application of cement is construction. The early Egyptians used to prepare the building material by mixing lime or clay and gypsum. Romans used to make the building material by mixing volcanic ash and lime. People all over Europe adopted the process developed by Romans for the purpose of construction. **Joseph Aspdin, a mason of Leeds, England** is the pioneer of the manufacture of cement. In the year, **1824**, he observed that a rock-like mass is obtained when water is added to a strongly heated mixture of limestone and clay. This hard mass resembles the Portland rock used by the people of England for the construction of building. Thus, the mixture was named as **Portland cement**.

Manufacture of Cement

The raw materials used for manufacturing Portland cement are limestone and clay. Cement can be produced in two different processes.

1. Dry process
2. Wet process

In the **dry process**, limestone and clay are used in definite proportion to achieve the desired property of the cement. The mixture is pulverised (making fine powder) using pulveriser and then heated to 1500–1600°C to form clinkers (nodules of mixed powder which becomes very hard after heating). These clinkers are again pulverised to form fine powder and small quantity of gypsum is added, and thus, cement is produced.

In the **wet process**, water is added to the pulverised mixture and slurry is obtained. Then the slurry is strongly heated to form clinkers.

Uses of Cement

Grouting: Grouting is a technique used in construction activities. Pure cement or cement mixed with sand and water in a definite proportion is applied in the gaps of construction material, like, bricks, tiles, etc., for making the surface smooth and strengthening the structure.

Cement mortar: It is a construction material prepared by mixing cement and sand in the ratio of 1 : 4 by weight. This mixture is converted into thick slurry by adding required quantity of water. It is used for joining bricks and plastering.

Concrete: It is also a construction material prepared by mixing cement, gravel and sand in the ratio of 1 : 2 : 3 by weight and required quantity of water. It is used for casting roofs, pillars, beams, etc.

Reinforced cement concrete: When cement concrete is reinforced using structural steel, it is called reinforced cement concrete (RCC). RCC structure provides huge strength and is durable. It is used for the construction of bridges, railway sleepers, prefabricated structures, columns, beams, roofs, etc.

Brick

Brick (red brick) is made by using clay. The right proportion of water is added to clay and kneaded properly. The kneaded clay is pressed in the shape of a brick and heated at high temperature (about 1000°C) and this process hardens the brick. Bricks are the structural materials used for construction of walls of a building.

Glass

Early cave people found an igneous rock named Obsidian which is a natural glass produced by volcanic action. They shaped it into tools and spearheads. With the advancement of science and technology, modern techniques have been adopted for the manufacture of a wide variety of glass.

Manufacture of Glass

Glass is made up of silica or sand, limestone and feldspar as major raw materials. Broken pieces of glass called cullet are added to the raw material. Some other ingredients, like, red lead, aluminium oxide, manganese oxide, chromium oxide, etc., are used to impart specific characteristics to glass. The raw materials are mixed in the required proportion and melted in a glass tank furnace or pot furnace. Cullet initiates the melting process at a lower temperature.

Three major techniques are applied on the molten glass obtained in the process of melting for the manufacture of different products of glass.

1. **Blowing of glass**: This technique is used for the manufacture of laboratory glassware, bottles, etc.
2. **Optical glass**: Molten glass is pressed into a circular shape from which different optical glasses are manufactured.
3. **Sheet glass/float glass**: Molten glass obtained from the glass tank furnace is placed on molten bath of tin. The molten glass spreads on the bath of molten tin with uniform thickness from which sheet glass/float glass is drawn.

The different shapes of molten glass obtained from the above three techniques are then subjected to a cooling process called annealing at controlled temperature in a long tunnel-type annealing chamber.

General Properties of Glass

Glass is a supercooled liquid. Molten glass obtained from glass manufacturing furnaces is allowed to cool down slowly. During cooling, viscosity of glass increases manifolds below the range of its freezing temperature and loses its ability to flow. Hence, it appears like a solid. Glass is transparent, highly chemical resistant, a bad conductor of heat and electricity and it can be easily moulded into desired shapes.

TABLE 6.5 Properties and applications of different types of glasses

Types of glasses	Ingredients and properties	Applications
Water glass	Sodium silicate, soluble in water.	1. It is used in the soap industry. 2. In the early 20th century, it was used for the preservation of eggs. 3. It is also used for filling up holes and strengthening material.
Soft glass or soda glass or lime glass	Sodium silicate and calcium silicate cannot withstand high temperature and thermal shock.	It is used for making window panes and cheap articles of glass.
Hard glass or potash–lime glass	Sand, limestone, potassium carbonate, fairly heat resistant.	It is used for making laboratory apparatus.
Optical glass or crown glass	Sand, limestone, potassium carbonate and red lead, highly transparent and possess high density.	It is used in the manufacture of lenses of spectacles and optical instruments, like, lenses of telescope, microscope, etc.
Jena glass or pyrex glass or borosil glass	Sand, limestone, sodium carbonate, aluminium oxide and borax, highly heat resistant and can withstand thermal shock.	It is used for making laboratory apparatus, thermos flasks and baking bowls.
Shatterpro of glass or sandwich glass	Sheets of plain glasses are attached with each other with a plastic sheet placed between them as broken pieces of this type of glass do not scatter.	It is used as wind shields of automobiles.

Polymers and Plastics

The single molecule present in the long chain is called monomer and the chain so formed by repetition of monomers is called polymer. One of the examples of such polymers is polyethylene. The monomer of this polymer is the ethylene molecule (C_2H_4). The word 'plastics' originated from the characteristic property of some materials to be moulded into various forms which is called plasticity. Plastics are made up of long chains called polymers obtained by the interlinkage of certain specific molecules of the same chemical composition. Plastics have many unique properties, like, light weight, mechanical strength, aesthetic appearance, bad conductor of heat and electricity, etc.

Based on the thermal behaviour, plastics can be divided into two categories: **thermoplastics** and **thermosetting plastics**.

Thermoplastics: This type of plastics can be processed repeatedly by melting and subsequent solidification without undergoing any chemical change. Hence, they can be reused. Examples of thermoplastics are polyethylene, polyvinyl chloride, polystyrene, nylon, terylene, etc.

Application of thermoplastics: Thermoplastics are used for making utensils, sole of the shoes, fabrics, etc.

Thermosetting plastics: This type of plastic cannot be softened by heating, and thus, cannot be reused.

Examples: Formaldehyde, polyurethane, melamine, bakelite, etc.

Application of thermosetting plastics: Thermosetting plastics are used for making water tanks, water pipes, cabinets of television and radio, unbreakable utensils, etc.

Soaps and Detergents

Soaps and detergents are the two artificially synthesized substances which are indispensable in our daily activity. Chemically soaps are the sodium salts of fatty acids. Examples of different fatty acids are olive oil, cotton seed oil, rapeseed oil, coconut oil, etc. When sodium hydroxide solution is treated with vegetable oil, soap is produced.

Vegetable oil + sodium hydroxide soap + glycerine. Glycerine is obtained as a by-product which has some different applications.

Kinds of Soaps

Various kinds of soaps and their applications are shown in Table 6.6.

TABLE 6.6 Properties and applications of different types of soaps

Kinds of soaps	Raw materials	Applications
Laundry soap or washing soap	Cheapest vegetable oil, contains free alkali.	It is used in the laundry for removing dirt and grease from soiled clothes.
Scouring soap	Finely ground pumice stone, sand and sodium carbonate are added to the laundry soap as fillers.	They are commonly used for scrubbing floors.
Toilet soap	The best quality of vegetable oil. Does not contain free alkali.	This kind of soap is used for bathing.
Medicated soap	Certain substances which have medicinal value are added to the toilet soap.	Neem soap takes care of the skin.
Soft soap or liquid soap	Vegetable oil and potassium hydroxide.	Used for washing hands. It is also used as a shaving soap.
Transparent soap	This kind of soap is obtained by dissolving soap in alcohol. When alcohol is evaporated, transparent soaps are formed.	This soap is also used as a bathing soap.

Detergents

Soap forms a curdy precipitate in hard water and does not lather properly. Hence, soap is ineffective in hard water. Detergents are sodium salts of alkyl hydrogen sulphate and can be used effectively in hard water. The cleansing action of the detergent is superior to that of the soap. Example of a detergent is sodium lauryl sulphate.

Cleansing Action of Soap and Detergent

Soaps and detergents remove the oil and greasy material which stick to the clothes. Soaps and detergents both are soluble in water and act as surfactant, i.e., they form a thin film which surrounds the oil and grease spots and help in the removal of those spots along with the dirt. As the removal of dirt by the soaps and detergents involve the formation of emulsion, it is called emulsification and these substances are called emulsifiers.

Paints

Paints are nothing but the colouring materials dissolved in the suspension of basic lead carbonate in linseed oil. Plastic emulsion is a kind of paint which is prepared by dissolving coloured plastics in amyl acetate. The durability of plastic emulsion is much more than that of the basic lead carbonate paints.

Medicines

In the olden days, people used to depend on different parts of medicinal plants, like, tulsi, vasaka, neem, etc., or their extracts for the treatment of diseases. Scientists observed and analysed the composition of those substances and developed similar kind of molecules artificially and the direct use of the medicinal plants has reduced gradually. With the advancement of the medical science, scientists developed many new molecules which are more effective and give quicker relief. Incurable diseases could also be cured by administering some of these molecules. These molecules are called medicines. Some of the examples of such medicines are penicillin, tetramycine, sulphadiazine, etc. Quinine is a medicine which is extracted from the bark of *Cinchona* plant. It is administered for the treatment of malaria.

EXAMPLE

A sample of hard water is taken in two containers A and B. To container A, a detergent is added and soap is added to container B. The water in A is more efficient in cleaning than the water in B. Justify.

SOLUTION

Effectiveness of soap is reduced when used in hard water. Hardness in water is caused by the presence of mineral salts mostly of Ca and Mg. The mineral salt reacts with soap to form an insoluble precipitate known as scum. Lather formation is reduced. Detergent forms soluble compounds with calcium and magnesium, and hence, does not form scum with hard water. Due to this reason, water in A is more efficient in cleaning than water in B.

EXAMPLE

Sand and clay which are abundantly available in the earth's crust have significant roles in the comfort zone of modern life. Justify.

SOLUTION

Sand is an important component in all types of mortars, such as, lime mortar and cement mortar. Sand is also the component of plain cement concrete and reinforced cement concrete (RCC) which contributes to the infrastructure development in various fields of modern life. Clay plays important role in ceramics, bricks, various building material, sanitary equipment and household articles. Because of their desired properties, they contribute a significant role in the comfort zone of modern life.

EXAMPLE

Write a short note on the cleansing action of detergent.

SOLUTION

Detergents remove the oil and greasy material which stick to clothes. Detergents are soluble in water and act as surfactants, that is, they from a thin film which surrounds the oil and grease spots and help in the removal of those spots along with the dirt. As the removal of dirt by soaps and detergents involves the formation of emulsion, it is called emulsification and these substances are called emulsifiers.

EXAMPLE

How are paints and plastic emulsions prepared?

SOLUTION

Paints are nothing but the colouring materials dissolved in the suspension of basic lead carbonate in linseed oil. Plastic emulsion is a kind of paint which is prepared by dissolving coloured plastics in amyl acetate. The durability of plastic emulsion is much more than that of basic lead carbonate paints.

EXAMPLE

What do you mean by grouting? Where does this technique find application?

SOLUTION

Grouting is a technique used in construction activities. Pure cement or cement mixed with sand and water in definite proportion is applied in the gaps of construction material, like, bricks, tiles, etc., for making the surface smooth and to strengthen the structure.

POINTS TO REMEMBER

- Aluminium is the third most abundant element in the earth's crust, whereas hydrogen is the most abundant element in the universe.
- Helium is the second lightest element and inert in nature, hence, it is used for filling weather observation balloons.
- Carbohydrates are instant sources of energy, whereas fats are reservoirs of energy.
- Methane is the simplest hydrocarbon and is present in natural gas or biogas as a major component.
- Polymers are large molecules formed by the combination of smaller molecules.
- Cement is a mixture of silicates and aluminates.
- Setting of cement is due to the hydration of these aluminates and silicates.
- Glass is a mixture of sodium silicate and calcium silicate. Addition of other components imparts other characteristic properties to glass, used for specific purposes.
- Glass is a supercooled liquid and appears like a solid due to the increase in viscosity during cooling of molten glass.
- Detergents are superior to soaps as they do not form insoluble precipitates with hard water.

TEST YOUR CONCEPTS

Very Short Answer Type Questions

Directions for questions from 1 to 10
Fill in the blanks.

1. Metals combine with other metals to form homogeneous solid mixture called _______.
2. The process of coating zinc on iron surface is called _______.
3. The alloy formed from gold and mercury is called _______.
4. _______ is used as a solid lubricant for machine parts.
5. Iodine in a mixture of potassium iodide and alcohol is called _______.
6. Zincite is obtained by heating _______.
7. The chemical formulae of plaster of Paris is _______.
8. The major component of natural gas is _______.
9. _______ is used for filling weather observation balloons.
10. Bakelite is an example of _______ plastic.

Directions for questions from 11 to 30
For each of the questions, four choices have been provided. Select the correct alternative.

11. The metals with the least tensile strength is
    (a) iron (b) aluminium
    (c) zinc (d) copper
12. Which of the following is not an alloy?
    (a) brass (b) bronz
    (c) magnalium (d) magnesium
13. Which of the following elements is used for tinning?
    (a) iron (b) tin
    (c) copper (d) Aluminium
14. The two metals involved in the galvanisation process are
    (a) iron and copper
    (b) iron and zinc
    (c) zinc and copper
    (d) copper and aluminium
15. Duralumin is used for the manufacture of aircraft frames because duralumin is
    (a) a bad conductor of heat and electricity.
    (b) more durable and has resistance to corrosion.
    (c) very light and possesses high tensile strength.
    (d) all of the above
16. Which of the following is a noble gas?
    (a) hydrogen (b) helium
    (c) oxygen (d) nitrogen
17. The elements present in stainless steel are
    (a) iron, copper and carbon
    (b) iron, copper and nickel
    (c) iron, copper and chromium
    (d) iron, chromium, carbon and nickel
18. Common salt is a mixture of
    (a) sodium chloride and magnesium chloride
    (b) sodium chloride and iodine
    (c) sodium chloride, iodine and potassium iodide
    (d) sodium chloride, sodium iodide or potassium iodide
19. The chemical composition of limestone is
    (a) calcium carbonate (b) calcium oxide
    (c) calcium hydroxide (d) calcium sulphate
20. The chemical name of sand is
    (a) silicon (b) sodium silicate
    (c) silicon dioxide (d) calcium silicate
21. Which among the following metals is the most suitable to be used as telecommunication cable?
    (a) aluminium (b) tin
    (c) iron (d) copper
22. Which of the following steps is not involved in the preparation of sodium chloride from sea water?
    (a) evaporation (b) dissolution in water
    (c) recryztalllisation (d) distillation
23. Which of the following properties of silicon is not exploited in the manufacture of non-stick cookware?

PRACTICE QUESTIONS

(a) high resistance to heat
(b) poor electrical conductivity
(c) non-stick nature
(d) high resistance to chemicals

24. Which of the following food items provide the maximum energy?
(a) vegetable oil (b) glucose
(c) cereals (d) pulses

25. Which of the following materials can be used for plastering?
(a) cement mortar (b) concrete
(c) RCC (d) anyone of the above

26. Which of the following is a difference between thermoplastic and thermosetting plastics?
(a) Thermoplastics can be processed repeatedly.
(b) Thermosetting plastics cannot be softened by heating.
(c) Thermoplastics do not undergo chemical change on heating.
(d) All the above.

27. Which of the following is obtained as a by-product in the manufacture of soap?
(a) glycerine (b) sodium hydroxide
(c) vegetable oil (d) salt of fatty acid

28. Glass in a/an
(a) amorphous solid
(b) crystalline solid
(c) supercooled liquid
(d) solid with a high viscosity

29. Arrange the compounds in sequence of increasing number of oxygen atoms:
(1) sugar (2) plaster of Paris
(3) starch (4) glucose
(a) 1243 (b) 2413
(c) 2314 (d) 2341

30. Arrange the steps in a sequence for the manufacture of cement in dry process.
(1) formation of clinker
(2) pulverisation
(3) addition of gypsum
(4) heating to high temperature (1500°C to 1600°C)
(a) 2413 (b) 2143
(c) 2341 (d) 2314

Directions for questions from 31 to 33
Match the entries given in Column A with the appropriate ones in Column B. More than one suitable content of Column B can also be considered correct for an entry in Column A and some content given in Column B may not be the appropriate one for any entry in Column A.

31.

Column A	Column B
A. Gold	( ) a. Sanitary pipes
B. Copper	( ) b. Satellite
C. Lead	( ) c. Covering tin roofs in cold countries
D. Zinc	( ) d. Bell metal
	( ) e. Iron roofing

32.

Column A	Column B
A. Surgical instruments	( ) a. Duralumin
B. Speed boats	( ) b. Bronze
C. Electrical switches	( ) c. Stainless steel
D. Household utensils	( ) d. Brass
	( ) e. Alloyed gold

33.

Column A	Column B
A. Water glass	( ) a. Potassium carbonate
B. Lime glass	( ) b. Sodium silicate
C. Crown glass	( ) c. Borax
D. Jena glass	( ) d. Calcium silicate
	( ) e. Aluminium oxide

Directions for questions from 34 to 53
Answer the following questions.

34. Mention some specific properties of metals.

35. Name two metals which find application in ayurvedic medicine.

36. Name the metal which is extensively used in a printed circuit board.

37. Why is graphite used as a major component in pencil lead?

38. Name one compound of silver which is used for cloud seeding.

39. Which element is used in the manufacture of insecticides and pesticides, like, DDT and BHC?

PRACTICE QUESTIONS

40. Why is magnesium used in fire works?

41. Which non-metal is used for the manufacture of pencil lead?

42. Why is duralumin used for the manufacture of automobiles and speed boats?

43. Which chemical compound is used to reduce the acidity of the soil?

44. Which compound of silicon is used for the preservation of eggs?

45. Name the hardest compound and write one of its use.

46. Write the formulae of glucose, sugar and starch.

47. Give an example of a solution which gives us instant energy.

48. Write the composition of stainless steel.

49. Name the soap commonly used for scrubbing floors.

50. How is concrete made?

51. What is emulsification?

52. Why is glucose given to patients?

53. What is annealing?

Short Answer Type Questions

54. How is sodium chloride extracted from sea water?

55. What is blue vitriol? Write its method of preparation and its uses.

56. How is plaster of Paris manufactured? Mention its applications.

57. How are bricks manufactured?

58. Name two different kinds of plastics and their properties with examples.

59. Distinguish soap from detergent.

60. Write a short note on medicines.

61. Write a short note on compounds of silicon.

62. Mention the physical properties of aluminium.

63. Discuss few applications of graphite.

64. Mention few properties of mercury.

65. Write a short note on marble.

66. Describe briefly the major techniques applied on the molten glass obtained during the manufacture of glass to get the desired variety of glass.

67. Discuss the general properties of glass.

68. Write a short note on hydrochloric acid.

Essay Type Questions

69. What are fertilisers? How are they classified? Explain the classification of fertilisers with an example.

70. Name three organic compounds which are used as food materials. Mention their constituents and uses.

71. How is cement manufactured? Mention the areas of application of cement.

72. How is glass manufactured? Mention different types of glass, their ingredients, properties and application.

73. What are the different kinds of soap? What are the raw materials used for the manufacture of different kinds of soap? Mention their applications.

74. (i) Name the metal which is used in the manufacture of high voltage electric transmission wire.
    (ii) Why is aluminium used in the manufacture of household articles?
    (iii) Write the name and the formula of the monomer of polyethylene.
    (iv) Among which alloys the percentage of zinc is less (brass/bronze)?
    (v) Why is silicone used in the manufacture of non-stick pans?

75. (i) What are hydrocarbons?
    (ii) Give the formula of the simplest hydrocarbon.

(iii) What is biogas?

(iv) Give two advantages of biogas.

(v) Name the products obtained on burning methane gas.

76. (i) Name the inert gas which is used for filling weather observation balloons.

(ii) Why is radon used in cancer treatment?

(iii) Why is an alloy of tin and lead used in the manufacture of fuse wire?

(iv) Name the gas which is used in advertising sign boards.

(v) Why is argon used in filling electric bulbs?

77. Mention the appropriate term corresponding to the following sentences.

(i) A solution obtained by passing carbon dioxide through sugar solution flavoured with desired substances.

(ii) A substance formed by highly concentrated sugar solution flavoured with desired substances.

(iii) A hard mass resembles the Portland rock used by the people for the construction of a building.

(iv) The hardest compound of silicon, used in all kinds of grinding wheels.

(v) A complex organic compound used as food, helps in development of cells, growth of the new cells and repair of damaged cells.

For Hints and Explanations, please visit: www.pearsoned.co.in/IITFoundationSeries

CONCEPT APPLICATION

Level 1

Directions for questions from 1 to 10
State whether the following statements are true or false.

1. Sodium is a very hard metal.
2. Non-metals are generally non-malleable and non-ductile.
3. Platinum is used as a catalyst.
4. Graphite is a good conductor of heat.
5. Xenon and krypton are used in electric bulbs to slow down the sublimation of tungsten.
6. Blue vitriol is used as a fungicide for grape plants.
7. Deficiency of potassium iodide causes goitre in humans.
8. Carbohydrates are the compounds of carbon, hydrogen and oxygen.
9. Water glass is used for making the lenses of telescopes.
10. Plastics are polymers.

Directions for questions from 11 to 20
Fill in the blanks.

11. _______ is an example of medicine extracted from the bark of a plant.
12. Duralumin contains _______ percentage of aluminium.
13. Noble gases are chemically _______.
14. _______ are generally non-malleable and non-ductile.
15. _______ is the least reactive metal which is used as a catalyst.
16. Xenon and krypton are used in electric bulbs to slow down the _______ of tungsten.
17. _______ are the organic compounds containing carbon, hydrogen, oxygen and nitrogen.
18. Graphite is a _______ conductor of heat.
19. _______ glass is used as wind shields of automobiles.
20. _______ is used as fuse wire in thermite welding.

Directions for questions from 21 to 41
For each of the questions, four choices have been provided. Select the correct alternative.

21. The elements present in starch are
    (a) carbon, hydrogen and sulphur
    (b) carbon, hydrogen and oxygen
    (c) carbon, hydrogen and nitrogen
    (d) carbon and hydrogen only
22. Pick out the correct statement(s) among the following:
    (a) An alloy is a homogeneous mixture of two or more metals.
    (b) An alloy may contain a non-metal in addition to metals.

PRACTICE QUESTIONS

(c) Alloys possess better physical properties than the metals.

(d) All the above.

23. Which among the following is a detergent?

(a) sodium stearate

(b) sodium chloride

(c) sodium carbonate

(d) sodium lauryl sulphate

24. Mortar is a mixture of

(a) cement and gravel

(b) gravel and sand

(c) cement, sand and water

(d) cement and water

25. Limestone is used in the manufacture of

(a) quick lime (b) glass

(c) cement (d) all of these

26. Iodised salt does not contain the following compound:

(a) sodium iodide (b) potassium iodide

(c) common salt (d) lithium iodide

27. Lead is used for making fusible alloy constituted of lead and tin in ______ ratio by mass.

(a) 1 : 2 (b) 2 : 1

(c) 1 : 3 (d) 1 : 1

28. Which among the following types of food can provide maximum amount of energy?

(a) starch (b) vegetable oil

(c) albumin (d) glucose

29. ______ is used as one of the raw materials for making soft soap.

(a) sodium carbonate

(b) alcohol

(c) potassium hydroxide

(d) finely ground pumice stone

30. The metallic compounds of which among the following metals is involved in cloud seeding?

(a) mercury (b) aluminium

(c) copper (d) silver

31. Which of the following applications of copper require refining of copper instead of alloying?

(a) making statues

(b) making household utensils

(c) making shells of ammunitions

(d) making electric transmission cables

32. Graphite is used in the manufacture of crucibles where a small quantity of metals can be melted. It is also used as electrodes. Which of the following properties of graphite is not exploited in these two given applications?

(a) highly resistance to chemicals

(b) soft nature

(c) high melting point

(d) good electrical conductivity

33.

Column A	Column B
A. Cleaning the surface of metal	( ) a. $CaSO_4$. ½ H_2O
B. Used as an insecticide	( ) b. $ZnCO_3$
C. Construction of Tajmahal	( ) c. HCl
D. Casting toys, statues, etc.	( ) d. $CuSO_4.5H_2O$
	( ) e. $CaCO_3$

(a) A – e, B – b, C – a, D – c

(b) A – b, B – c, C – a, D – e

(c) A – c, B – d, C – e, D – a

(d) A – d, B – c, C – a, D – b

34. Which of the following can be said to provide instant energy?

(a) A solution of glucose (b) A cup of tea

(c) Aerated drinks (d) Milk

35. **Assertion (A)**: Fuse wire is made up of an alloy constituted of lead and tin in 1 : 1 ratio by mass.

Reason (R): Both lead and tin have comparatively low melting points than other metals.

(a) Both (A) and (R) are true and (R) is the correct reason of (A).

(b) Both (A) and (R) are true but (R) is not the correct explanation of (A).

(c) (A) is true but (R) is false.

(d) (A) is false but (R) is true.

36. Which of the following is the most suitable for the construction of a pillar?

(a) cement mortar (b) concrete

(c) RCC (d) pure cement

37. Which of the following is a thermosetting plastic?

(a) polyethylene (b) nylon

(c) terelyne (d) bakelite

38. Which of the following substances does not form a curdy precipitate when it is added to hard water?
(a) potassium salt of fatty acid
(b) sodium salt of fatty acid
(c) sodium lauryl sulphate
(d) none of the above

39. **Assertion (A)**: Gold is used for making jewellery coins.
Reason (R): It is relatively a hard metal.
(a) Both (A) and (R) are true and (R) is the correct reason of (A).
(b) Both (A) and (R) are true but (R) is not the correct explanation of (A).
(c) (A) is true but (R) is false.
(d) (A) is false but (R) is true.

40. Arrange the following steps in a sequence for the manufacture of glass.
(1) addition of cullet
(2) mixing of raw materials in a proper proportion
(3) addition of colouring pigments
(4) annealing
(a) 2134 (b) 2341
(c) 3214 (d) 2431

41. Arrange the following steps in a sequence for the manufacture of common salt.
(1) recrystallisation
(2) evaporation of sea water
(3) addition of sodium iodide and potassium iodide
(4) dissolution in water
(5) formation of crude salt
(a) 25413 (b) 24513
(c) 23415 (d) 12543

Directions for questions from 42 to 44

Match the entries given in Column A with the appropriate ones in Column B. More than one suitable content of Column B can also be considered correct for an entry in Column A and some content given in Column B may not be the appropriate one for any entry in Column A.

42.

Column A	Column B
A. Silver	( ) a. Jewellery
B. Aluminium	( ) b. Sterilisation of water
C. Gold	( ) c. Artificial rain
D. Chlorine	( ) d. Dentistry
	( ) e. Magnalium

43.

Column A	Column B
A. Duralumin	( ) a. Making statues
B. Brass	( ) b. Imitation jewellery
C. Alloyed gold	( ) c. Pressure cooker
D. Bronze	( ) d. Ammunition
	( ) e. Aircraft body

44.

Column A	Column B
A. Cement mortar	( ) a. Grouting
B. Concrete	( ) b. Prefabricated structures
C. Reinforced cement concrete	( ) c. Plastering
D. Pure cement	( ) d. Cement, gravel and sand
	( ) e. Used for joining bricks

Level 2

Directions for questions from 1 to 12

For each of the questions, four choices have been provided. Select the correct alternative.

1. Arrange the following steps in a sequence for the manufacture of cement.
(1) Mixture is pulverised using a pulveriser.
(2) Limestone and clay are mixed in a definite proportion.
(3) Mixture is heated to 1500°C to 1600°C to form clinkers.
(4) Clinkers are pulverised to form fine powder and a small quantity of gypsum is added.
(a) 1324 (b) 2134
(c) 2431 (d) 4321

2. Which of the following metals cannot be used in printed circuit boards?
(a) copper (b) aluminium
(c) silver (d) zinc

3. Graphite is used as a pencil lead because
(a) it has a high melting point
(b) it is a good conductor of heat and electricity
(c) it is soft in nature
(d) it is highly resistant to chemicals

PRACTICE QUESTIONS

4. Which of the following gases is the most unsuitable for filling up weather observation balloons?
   (a) helium (b) neon
   (c) hydrogen (d) argon

5. Which of the following fertilisers cannot replenish nitrogen?
   (a) superphosphate of lime (calcium dihydrogen phosphate with gypsum)
   (b) nitre
   (c) ammonium sulphate
   (d) triammonium phosphate
   (a) A – c, B – d, C – b, D – a
   (b) A – e, B – a, C – b, D – c
   (c) A – b, B – e, C – a, D – c
   (d) A – d, B – a, C – b, D – e

6.

Column A	Column B
A. Hydrochloric acid	( ) a. Sprayed over grape vines
B. Hydrated copper sulphate	( ) b. Flooring, statues and monuments
C. Limestone	( ) c. Manufacture of paints
D. Zincite	( ) d. Reduces the acidity of the soil
	( ) e. Bathroom cleaning agent

   (a) A – c, B – d, C – b, D – a
   (b) A – e, B – a, C – b, D – c
   (c) A – b, B – e, C – a, D – c
   (d) A – d, B – a, C – b, D – e

7. **Assertion (A)**: Tin is used in the manufacture of fusible alloys.
   Reasons (R): Tin is a good conductor of electricity.
   (a) Both (A) and (R) are true and (R) is the correct reason of (A).
   (b) Both (A) and (R) are true but (R) is not the correct explanation of (A).
   (c) (A) is true but (R) is false.
   (d) (A) is false but (R) is true.

8. Arrange the following steps in a sequence for the manufacture of borosilicate glass.
   (1) addition of ingredients, like, red lead, aluminium oxide, manganese oxide, chromium oxide, etc.
   (2) addition of broken pieces of glass.
   (3) mixing up of sand, limestone, sodium carbonate, aluminium oxide and borax in required proportions.
   (4) melting mixture in a glass tank furnace or pot furnace.
   (5) annealing process
   (6) blowing of glass
   (a) 321465 (b) 321654
   (c) 342165 (d) 365214

9. Which among the following metals is the most suitable to be used as telecommunication cable?
   (a) aluminium (b) tin
   (c) iron (d) copper

10. Arrange the suitable steps in a correct sequence for the manufacture of sodium chloride from sea water.
   (1) making of hot saturated solution
   (2) evaporation
   (3) recrystallisation
   (4) collection of sea water
   (5) addition of a small quantity of sodium iodide or potassium iodide
   (a) 4521 (b) 4523
   (c) 45123 (d) 4213

11. Which of the following noble gases cannot be used for filling up electric bulbs?
   (a) helium (b) neon
   (c) argon (d) radon

12. Which of the following compounds/alloys of silicon is used in grinding wheel and which property of that substance is exploited?
   (a) silicon carbide because it is the hardest substance
   (b) ferrosilicon because it is resistant to friction
   (c) sodium silicate because it becomes impervious when heated strongly
   (d) silicon dioxide because it is resistant to chemicals

13. In an 'aqua regia' quiz competition the quiz master asked one of the teams to name the two hardest substances one of which is naturally occurring and the other one is manmade. As none of the teams could give the answer, he gave a clue that both the substances are used for cutting glass and drilling rocks and contain one common

element. Then one of the teams gave the answer. What could be the right answer?

14. Though fat provides the maximum amount of energy, it forms a minor part of our diet while a large part of diet contains carbohydrates?

15. An apartment owner called for a plumber to replace one of the leaking pipe lines. The plumber arrived and found that the galvanised iron pipes are deteriorated and corroded. He advised the apartment owner to replace the galvanised iron pipes with the PVC pipes. Why did the plumber give the above suggestion?

16. A chemistry teacher was teaching polymers in a class. He said that though bakelite and polyethylene belong to polymer family, only bakelite is used for making electrical insulators or handles of kitchenware. Why cannot polyethylene be used for the above purpose?

17. In a chemistry laboratory, a research scholar while conducting an experiment, observed at a certain step, that a yellow-coloured gas (X) was released and it has been found to be poisonous. This was later made to react with a highly reactive metal (Y). A new compound (Z) was formed which is used for preservation of fish, meat and pickles. The professor asked him to identify X, Y and Z.

18. Three naturally occurring elements X, Y and Z are good conductors of electricity and are very precious. Z is more reactive than X and Y and on exposure to air for a long time turns black. Z forms an alloy with Y and is used in jewellery and X is used as a catalyst. Identify X, Y and Z.

19. A reddish brown coloured metal P and a silvery white coloured metal Q are highly ductile and are good conductors of heat and electricity. P plays a major role in the preparation of brass and bronze, whereas Q is the most abundant element in the earth's crust, after oxygen and silicon, but Q is used as a substitute for P. Give reasons.

20. A gaseous element X is used in the manufacture of bleaching powder. On reaction with hydrogen X forms a substance Y, which after dissolution in water is used to clean metal surfaces. Identify X and Y. Justify the whole process involved.

21. An element X is non-malleable and non-ductile. X is extensively used to protect another element Y from corrosion. Identify X and Y and name the process that is mentioned above. Also, compare the reactivities of X and Y.

22. A salt X of a metal which is used in the galvanising process on heating produces an oxide of that metal Y and a gas Z. Z is used in the preparation of soda water, and Y is used as a base of white paint. Identify X, Y and Z.

23. A non-metallic element X is a good conductor of electricity in one form and is a bad conductor of electricity in another form. Comment on the thermal conductivity of both the forms. Identify the element and mention the above two forms of X.

24. A salt A on heating forms the oxide of the metal B and a gas C. B is used in white washing. When C is passed through water, it forms D. Identify A, B, C and D. Justify the usage of C for the above purpose.

Level 3

1. Cement replaces lime mortar in the field of construction. Explain.

2. People used to apply a coating of sodium silicate on eggs for the purpose of preservation. Justify.

3. Name the metal which is extensively used in thermometers. Justify the reason for the same.

4. Pyrex or borosil glass is used for making laboratory apparatus, whereas crown glass is used in the manufacture of optical instruments. Give a reason for the same.

5. A non-metal X is lustrous, insoluble in water but soluble in alcohol, when it reacts with another metallic element Y, it forms a substance Z which are used in cloud seeding or artificial rain. Identify X, Y and Z. What are the advantages of artificial rain? Is there any substances other than Z that can be used for artificial rain? If so, what are they?

PRACTICE QUESTIONS

ASSESSMENT TEST

Test 1

Directions for questions from 1 to 15

Select the correct alternative from the given choices.

1. Which of the following properties of silicon is not exploited in the manufacture of non-stick cookware?
   (a) high resistance to heat
   (b) poor electrical conductivity
   (c) non-stick nature
   (d) high resistance to chemicals

2. Which of the following is a difference between thermoplastic and thermosetting plastics?
   (a) Thermoplastics can be processed repeatedly.
   (b) Thermosetting plastics cannot be softened by heating.
   (c) Thermoplastics do not undergo chemical change on heating.
   (d) All the above

3. Arrange the given compounds in the increasing order of carbon atoms present in the following compounds:
   (1) methane (2) sugar
   (3) glucose (4) starch
   (a) 2 3 4 1 (b) 3 1 2 4
   (c) 1 3 2 4 (d) 4 2 1 3

4. **Assertion (A)**: Gold threads are used in silk sarees.
   Reason (R): Gold is highly ductile by nature.
   (a) Both (A) and (R) are true and (R) is the correct reason of (A).
   (b) Both (A) and (R) are true but (R) is not the correct explanation of (A).
   (c) (A) is true but (R) is false.
   (d) (A) is false but (R) is true.

5. Which of the following gases is the most unsuitable for filling up weather observation balloons?
   (a) helium (b) neon
   (c) hydrogen (d) argon
   (a) A → c, B → d, C → b, D → a
   (b) A → e, B → a, C → b, D → c
   (c) A → b, B → e, C → a, D → c
   (d) A → d, B → a, C → b, D → e

6. Match the following:

Column A	Column B
A. Hydrochloric acid	( ) a. Sprayed over grape vines
B. Hydrated copper sulphate	( ) b. Flooring, statues and monuments
C. Limestone	( ) c. Manufacture of paints
D. Zincite	( ) d. Reduces the acidity of the soil
	( ) e. Bathroom cleaning agent

   (a) A → c, B → d, C → b, D → a
   (b) A → e, B → a, C → b, D → c
   (c) A → b, B → e, C → a, D → c
   (d) A → d, B → a, C → b, D → e

7. Which of the following is obtained as a by-product in the manufacture of soap?
   (a) glycerine (b) sodium hydroxide
   (c) vegetable oil (d) salt of fatty acid

8. Which of the following metals cannot be used in printed circuit boards?
   (a) copper (b) aluminium
   (c) silver (d) zinc

9. Which of the following food items provides the maximum energy?
   (a) vegetable oil (b) glucose
   (c) cereals (d) pulses

10. Copper is mixed with gold to make ornaments because alloying
    (a) increases the hardness of gold
    (b) increases the malleability of gold
    (c) makes gold a very good conductor of heat
    (d) decreases its fusibility

11. Which of the following fertilisers replenishes only one essential nutrient among nitrogen, phosphorus and potassium?
    (a) potassium nitrate
    (b) triammonium phosphate
    (c) potassium bisulphate
    (d) all of these

12. Arrange the following steps in a sequence for the manufacture of cement.
   (1) Mixture is pulverised using a pulveriser.
   (2) Limestone and clay are mixed in a definite proportion.
   (3) Mixture is heated to 1500°C to 1600°C to form clinkers.
   (4) Clinkers are pulverised to form fine powder and a small quantity of gypsum is added.
   (a) 1 3 2 4 (b) 2 1 3 4
   (c) 2 4 3 1 (d) 4 3 2 1

13. Graphite is used as a pencil lead because
   (a) it has a high melting point
   (b) it is a good conductor of heat and electricity
   (c) it is soft
   (d) it is highly resistant to chemicals

14. Which of the following materials can be used for plastering?
   (a) cement mortar (b) concrete
   (c) RCC (d) anyone of these

15. **Assertion (A)**: Tin is used in the manufacture of fusible alloys.
   Reasons (R): Tin is a good conductor of electricity.
   (a) Both (A) and (R) are true and (R) is the correct reason of (A).
   (b) Both (A) and (R) are true but (R) is not the correct explanation of (A).
   (c) (A) is true but (R) is false.
   (d) (A) is false but (R) is true.

Test 2

Directions for questions from 1 to 15

Select the correct alternative from the given choices.

1. Which of the following compounds/alloys of silicon is used in grinding wheel and which property of that substance is exploited?
   (a) silicon carbide because it is the hardest substance
   (b) ferrosilicon because it is resistant to friction
   (c) sodium silicate because it becomes impervious when heated strongly
   (d) silicon dioxide because it is resistant to chemicals

2. Which of the following is a thermosetting plastic?
   (a) polyethylene (b) nylon
   (c) terelyne (d) bakelite

3. Arrange the relevant steps in correct sequence for the manufacture of sodium chloride from sea water.
   (1) making of hot saturated solution
   (2) evaporation
   (3) recrystallisation
   (4) collection of sea water
   (5) addition of small quantity of sodium iodide or potassium iodide
   (a) 4 5 2 1 (b) 4 5 2 3
   (c) 4 5 1 2 3 (d) 4 2 1 3

4. **Assertion (A)**: Gold is used for making jewellery coins.
   Reason (R): It is relatively a hard metal.
   (a) Both (A) and (R) are true and (R) is the correct reason of (A).
   (b) Both (A) and (R) are true but (R) is not the correct explanation of (A).
   (c) (A) is true but (R) is false.
   (d) (A) is false but (R) is true.

5. Which of the following noble gases cannot be used for filling up electric bulbs?
   (a) helium (b) neon
   (c) argon (d) radon

6. Match the following:

Column A	Column B
A. Cleaning the surface of metal	( ) a. $CaSO_4.\frac{1}{2}H_2O$
B. Used as an insecticide	( ) b. $ZnCO_3$
C. Construction of Taj Mahal	( ) c. HCl
D. Casting toys, statues, etc.	( ) d. $CuSO_4.5H_2O$
	( ) e. $CaCO_3$

   (a) A → e, B → b, C → a, D → c
   (b) A → b, B → c, C → a, D → e
   (c) A → c, B → d, C → e, D → a
   (d) A → d, B → c, C → a, D → b

PRACTICE QUESTIONS

7. Which of the following substances does not form a curdy precipitate when it is added to hard water?
   (a) potassium salt of fatty acid
   (b) sodium salt of fatty acid
   (c) sodium lauryl sulphate
   (d) none of these

8. Which among the following metals is the most suitable to be used as telecommunication cable?
   (a) aluminium (b) tin
   (c) iron (d) copper

9. Which of the following can be said to provide instant energy?
   (a) a solution of glucose (b) a cup of tea
   (c) aerated drinks (d) milk

10. Which of the following applications of copper require refining of copper instead of alloying?
   (a) making statues
   (b) making household utensils
   (c) making shells of ammunitions
   (d) making electric transmission cables

11. Which of the following fertilisers cannot replenish nitrogen?
   (a) superphosphate of lime (calcium dihydrogen phosphate with gypsum)
   (b) nitre
   (c) ammonium sulphate
   (d) triammonium phosphate

12. Arrange the following steps in a sequence for the manufacture of borosilicate glass.
   (1) addition of ingredients like red lead, aluminium oxide, manganese oxide, chromium oxide, etc.
   (2) addition of broken pieces of glass
   (3) mixing up of sand, limestone, sodium carbonate, aluminium oxide and borax in required proportions
   (4) melting mixture in a glass tank furnace or pot furnace
   (5) annealing process
   (6) blowing of glass

   (a) 3 2 1 4 6 5 (b) 3 2 1 6 5 4
   (c) 3 4 2 1 6 5 (d) 3 6 5 2 1 4

13. Graphite is used in the manufacture of crucibles where a small quantity of metals can be melted. It is also used as electrodes. Which of the following properties of graphite is not exploited in these two given applications?
   (a) highly resistant to chemicals
   (b) soft nature
   (c) high melting point
   (d) good electrical conductivity

14. Which of the following is the most suitable for the construction of a pillar?
   (a) cement mortar (b) concrete
   (c) RCC (d) pure cement

15. **Assertion (A)**: Fuse wire is made up of an alloy constituted of lead and tin in 1 : 1 ratio by mass.
   Reason (R): Both lead and tin have comparatively lower melting points than other metals.
   (a) Both (A) and (R) are true and (R) is the correct reason of (A).
   (b) Both (A) and (R) are true but (R) is not the correct explanation of (A).
   (c) (A) is true but (R) is false.
   (d) (A) is false but (R) is true.

TEST YOUR CONCEPTS

Fill in the blanks

1. Alloys
2. Galvanisation
3. Gold amalgam
4. Graphite
5. Tincture of iodine
6. Zinc carbonate
7. $CaSO_4.1/2\ H_2O$
8. Methane
9. Helium
10. Thermosetting

Multiple choice questions

11. c	12. d	13. b	14. b	15. c
16. b	17. d	18. d	19. a	20. c
21. d	22. d	23. b	24. a	25. a
26. d	27. a	28. c	29. b	30. a

Match the following

31. A → b
B → d
C → a, c
D → e

32. A → c
B → a
C → b
D → c, d, b

33. A → b
B → b, d
C → a
D → c, e

CONCEPT APPLICATION

Level 1

True or False

1. False	2. True	3. True	4. False	5. True
6. True	7. False	8. True	9. False	10. True

Fill in the blanks

11. Quinine
12. 95
13. Inert
14. Non-metals
15. Platinum
16. Sublimation
17. Proteins
18. Bad
19. Sandwitch glass
20. Magnesium

Multiple choice questions

21. b	22. d	23. d	24. c	25. d
26. d	27. d	28. b	29. c	30. d
31. d	32. b	33. c	34. a	35. a
36. c	37. d	38. c	39. c	40. a
41. a				

Match the following

42. A → c, a, d
B → e
C → a, d
D → b

43. A → c, e
B → a
C → b
D → d

44. A → c, e
B → d
C → b
D → a

Level 2

Multiple choice questions

1. b	2. d	3. c	4. d	5. a
6. b	7. b	8. a	9. d	10. d
11. d	12. a			

ASSESSMENT TEST

Test 1

1. b	2. d	3. c	4. a	5. d
6. b	7. a	8. d	9. a	10. a
11. c	12. b	13. c	14. a	15. b

Test 2

1. a	2. d	3. d	4. c	5. d
6. c	7. c	8. d	9. a	10. d
11. a	12. a	13. b	14. c	15. a

CONCEPT APPLICATION

Level 1

True or false

1. Reason: Sodium is a very soft metal.
2. Reason: Non-metals are generally non-malleable and non-ductile
3. Reason: Platinum used as a catalyst for most of the reactions that take place in gaseous state.
4. Reason: Graphite is a good conductor of electricity but bad conductor of heat.
5. Reason: Xenon and kryptons are used to slow down the sublimation of tungsten in a bulb.
6. Reason: Blue vitriol is used as a fungicide for fruit trees in the form of Bordeaux mixture.
7. Reason: Deficiency of iodine causes goitre in human beings.
8. Reason: Carbohydrates generally contain carbon, hydrogen and oxygen.
9. Reason: Optical glass is used for making the lenses of telescopes.
10. Reason: Plastics are polymers.

Fill in the blanks

11. Quinine is an example of medicine extracted from the bark of a plant.
12. Duralumin contains 95% of aluminium.
13. Noble gases are chemically inert.
14. Non-metals are generally non-malleable and non-ductile.
15. Platinum is the least reactive metal which is used as a catalyst.
16. Xenon and krypton are used in electric bulbs to slow down the sublimation of tungsten.
17. Proteins are the organic compounds containing carbon, hydrogen, oxygen and nitrogen.
18. Graphite is a bad conductor of heat.
19. Sandwitch glass is used as wind shields of automobiles.
20. Magnesium is used as fuse wires in thermite welding.

Multiple choice questions

21. Starch is an organic compound used as a food having the formula $(C_6H_{10}O_5)_n$ and consists of carbon, hydrogen and oxygen.
22. An alloy is a homogeneous mixture of two or more metals or may contain a non-metal in addition to metals. These alloys possess better physical properties than the metals.
23. Example of detergent is sodium lauryl sulphate.
24. Mortar is a mixture composed of cement, sand and water.
25. Lime stone is one of the raw materials used for the manufacture of quicklime, glass and cement.
26. The common salt containing sodium iodide and potassium iodide is called iodised salt. Lithium iodide is not present in the iodised salt.
27. Lead is used for making fusible alloy constituted lead and tin in 1 : 1 ratio by mass.
28. Fats provide maximum amount of energy and vegetable oil is an example of fat.
29. Potassium hydroxide is used as one of the raw materials for making soft soap.
30. Silver nitrite or silver iodide is used in cloud seeding for artificial rain.
31. Pure copper is required to manufacture electric transmission cables.
32. Graphite is soft but this property is not exploited in the given applications.
33. A → c Hydrochloric acid is used in cleaning the surface of metal.

    B → d Hydrated copper sulphate is used as an insecticide.

C → e Taj Mahal which is one of the seven wonders is constructed with marble.

D → a Plaster of Paris is used for casting statues, toys and decorative materials.

34. A solution of glucose, which is the simplest carbohydrate, provides instant energy.

35. Fuse wire is made up of an alloy constituted of lead and tin in 1 : 1 ratio by mass as both lead and tin have comparatively low melting points than other metals.

36. RCC is very strong because of the presence of structural steel inside it. Hence, it is most suitable for the construction of pillars.

37. Bakelite is a thermosetting plastic

38. Sodium lauryl sulphate does not form a curdy precipitate in hard water.

39. Gold is used for making jewellery coins as gold possesses a yellow lusture, is malleable, ductile, soft and highly resistant to chemicals.

40. (i) mixing of raw materials in proper proportion
(ii) addition of cullet
(iii) addition of colour pigments
(iv) annealing

41. (i) evaporation of sea water
(ii) formation of crude salt
(iii) dissolution in water
(iv) recrystallisation
(v) addition of sodium iodide and potassium iodide

Match the following

42. A → c, a, d Silver compounds are useful in forcing artificial rain, in dentistry and in making jewellery.

B → e Magnalium is an alloy of aluminium.

C → a, d Gold is used in making jewellery and dentistry

D → b Chlorine is used in the sterlisation of water.

43. A → c, e Duralumin is an alloy used in making pressure cooker and aircraft body.

B → a Brass is an alloy of copper used in making statues.

C → b Alloyed gold is used in making imitation jewellery.

D → d Bronze alloy is used in making ammunition and statues.

44. A → c, e Cement mortar is used in plastering and for joining bricks.

B → d Concrete is a mixture of cement, gravel and sand.

C → b Reinforced cement concrete is used for making prefabricated structures.

D → a Pure cement is used in grouting.

Level 2

Multiple choice questions

1. (i) Limestone and clay are mixed in a definite proportion.
(ii) Mixture is pulverised using a pulveriser.
(iii) Mixture is heated to 1500°C to 1600°C to form clinkers.
(iv) Clinkers are pulverised to form fine powder and a small quantity of gypsum is added.

2. Zinc is brittle, and hence, cannot be used in printed circuit boards.

3. Graphite is used as pencil lead because it is soft and leaves an impression on the paper.

4. Argon is unsuitable for filling up weather observation balloons because it is heavier than air.

5. Superphosphate of lime cannot provide nitrogen.

6. A – e Hydrochloric acid is used as a bathroom cleaning agent.

B – a Hydrated copper sulphate is sprayed over grape vines to protect them from insects.

C – b Limestone is extensively used in flooring, statues and monuments.

D – c Zincite is used for the manufacture of paints.

7. Tin is used in the manufacture of fusible alloy as it has very low melting point.

8. (i) mixing up of sand, limestone, sodium carbonate, aluminium oxide and borax in required proportion.

(ii) addition of broken pieces of glass.

(iii) addition of ingredients, like, red lead, aluminium oxide, manganese oxide, chromium oxide, etc.

(iv) melting mixture in a glass tank furnace or pot furnace.

(v) blowing of glass.

(vi) annealing process.

9. Electrical conductivity of copper is the maximum.

10. (i) collection of sea water

(ii) evaporation

(iii) making of hot saturated solution

(iv) recrystallisation

11. Radon, being radioactive, cannot be used to fill up electric bulbs.

12. Silicon carbide is used in grinding wheels because it is the hardest substance.

13. Diamond and silicon carbide are the two hardest substances known. Diamond occurs naturally (one is diamond and the other is silicon carbide) as it is obtained by the reaction between carbon and silicon. The common element present in diamond and silicon carbide is carbon.

14. Both fats and carbohydrates are energy giving food nutrients and both contain carbon, hydrogen and oxygen, but the molecular structure of fat is more complex than that of carbohydrates. Carbohydrates break up to glucose which are simple molecules and which on combustion in presence of oxygen breaks down to form carbon dioxide and water and releases energy. Combustion of fat takes a long period of time, and hence, carbohydrates are preferable to fats. Fats are considered as reservoirs of energy while carbohydrates serve as an instant source of energy.

15. PVC pipes are polyvinyl chloride pipes which are thermoplastic in nature and more durable as they are completely rust free and less expensive than galvanised iron pipes. Though galvanisation prevents rusting, since the pipelines are in continuous contact with water, the coating may get worn off over a period of time and it may be corroded.

16. Polythene and bakelite are the polymers and have a lot of applications in daily life. Since polythene is a thermoplastic (which melts on heating), therefore, cannot be used for making electrical insulators or handles. Bakelite can be used for these purposes as it is a thermosetting polymer (once set into hard mass cannot be remelted).

17. A greenish yellow poisonous gas X is chlorine. On reaction with the metal sodium (highly reactive metal) chlorine gas forms sodium chloride which is used for preservation of fish, meats and pickles. So, Z is sodium chloride, X is chlorine gas and Y is sodium metal.

18. Three precious elements that occur in nature are silver, gold and platinum. Silver is more reactive than gold and platinum. On exposure to air for longer time, silver turns black due to the formation of silver sulphide. Platinum is used as a catalyst. Alloy of gold with silver is used in jewellery. So, X is platinum, Y is gold and Z is silver.

19. Brass and bronze mainly contain copper which is a reddish brown coloured metal. The most abundant element in the earth's crust after oxygen and silicon is aluminium which is silvery white coloured metal. So, P is copper and Q is aluminium. Both copper and aluminium are good conductors of heat and electricity. Since copper is a costly metal, aluminium is used as a substitute for copper.

20. The element chlorine is used in the manufacture of bleaching power. It reacts with hydrogen to form hydrogen chloride which dissolves in water to give hydrochloric acid. It is used to clean the metal surfaces. So, X is chlorine and Y is hydrochloride which on dissolution in water gives hydrochloric acid. Metal surfaces have coating of metal oxides and hydrochloric acid can react with the metal oxides and removes the coating.

21. Element X is zinc which is non-malleable and non-ductile. Y is iron. The process of protecting iron by zinc is called galvanisation. The element X(Zn) is more reactive than Y (Fe).

22. Since Z is used in making soda water, it is carbon dioxide. The metal carbonate is zinc carbonate

because zinc carbonate on heating produces zinc oxide and carbon dioxide. Zinc oxide is used as paint. So, X is zinc carbonate, Y is zinc oxide and Z is carbon dioxide.

23. Carbon occurs as graphite and diamond in a free state. Graphite is a good conductor of electricity and diamond is a bad conductor of electricity. So, the element X is carbon and the two forms of X are graphite and diamond. Diamond is a good thermal conductor, whereas graphite is a poor thermal conductor.

24. Metal carbonate is calcium carbonate which on heating forms calcium oxide and carbon dioxide. Calcium oxide is added to water to form calcium hydroxide which is used for white washing. Carbon dioxide forms carbonic acid (a weak acid) when dissolved in water and turns lime water milky when passed through it. So, A is calcium carbonate, B is calcium oxide, C is carbon dioxide and D is carbonic acid. When white washing of walls is done, calcium hydroxide reacts with atmospheric CO_2 to form a white powdery coating of calcium carbonate.

Level 3

1. Both cement mortar and lime mortar are used for construction since they get hardened on exposure to air. But cement is a stronger binding material than lime and due to complex chemical reactions it forms a harder mass over a period of time. Moreover, on combining with gravel, it forms concrete which is much harder. Its strength is further enhanced when used in combination with steel in the form of RCC. Therefore, cement replaces lime mortar as a better construction material.

2. Sodium or potassium silicate is used as a protective covering for eggs. When egg is dipped in a solution of sodium silicate in water, a layer of sodium silicate forms on the surface of the egg which is very impervious to water, air and heat. As a result, bacterial growth is prevented in the egg so that it can be preserved for relatively longer periods of time.

3. Mercury is generally used in clinical thermometers as it expands on heating and has high density and does not stick to container.

4. Pyrex or borosil glass consists of borax which reduces the expandability of glass and provides heat resistance, whereas presence of red lead in crown glass increases the density of glass to a great extent making it useful for optical instruments.

5. A lustrous non-metal which is soluble in alcohol is iodine. It reacts with the metallic element silver to form silver iodide which is used in cloud seeding or artificial rain. So, X is iodine, Y is silver, Z is silver iodide. Artificial rain is very much useful in agriculture at the time of drought or water scarcity. Silver nitrate and dry ice (solid carbon dioxide) can also be used for artificial rain.

ASSESSMENT TEST

Test 1

1. Silicone is a bad conductor of electricity and this property is not exploited in the manufacture of non-stick cookware.

2. Thermoplastic can be processed repeatedly, can be softened by heating and do not undergo any chemical change on heating.

3. (a) Methane (b) Sugar
(c) Glucose (d) Starch

4. In silk sarees gold threads could be used as gold is highly ductile by nature.

5. Argon is unsuitable for filling up weather observation balloons because it is heavier than air.

6. A → e: Hydrochloric acid is used as a bath room cleaning agent.
B → a: Hydrated copper sulphate is sprayed over grape vines to protect them from insects.

C → b: Limestone is extensively used for flooring, statues and monuments.

D → c: Zincite is used for the manufacture of paints.

7. Glycerine is obtained as a by-product in the manufacture of soap.

8. Zinc is brittle, and hence, cannot be used in printed circuit boards.

9. Vegetable oil is a fat and it provides the maximum energy.

10. Copper is mixed with gold to make ornaments in order to increase its hardness.

11. Potassium bisulphate replenishes only potassium to the soil.

12. (a) Limestone and clay are mixed in a definite proportion.
(b) Mixture is pulverised using a pulveriser.
(c) Mixture is heated to 1500°C–1600°C to form clinkers.
(d) Clinkers are pulverised to form fine powder and a small quantity of gypsum is added.

13. Graphite is used as pencil lead because it is soft and leaves an impression on the paper.

14. Cement mortar is used for plastering.

15. Tin is used in the manufacture of fusible alloy as it has a very low melting point.

Test 2

1. Silicon carbide is used in grinding wheels because it is the hardest substance.

2. Bakelite is a thermosetting plastic.

3. (d) Collection of sea water
(b) Evaporation
(a) Making of hot saturated solution
(c) Recrystallisation

4. Gold is used for making jewellery coins as gold possesses a yellow lustre, is malleable, ductile, soft and highly resistant to chemicals.

5. Radon, being radioactive, cannot be used to fill up electric bulbs.

6. A → c: Hydrochloric acid is used in cleaning the surface of metals.

B → d: Hydrated copper sulphate is used as an insecticide.

C → e: Taj Mahal which is one of the seven wonders is constructed with marble.

D → a: Plaster of Paris is used for casting statues, toys and decorative materials.

7. Sodium lauryl sulphate does not form a curdy precipitate in hard water.

8. Electrical conductivity of copper is the maximum.

9. A solution of glucose, which is the simplest carbohydrate, provides instant energy.

10. Pure copper is required to manufacture electric transmission cables.

11. Superphosphate of lime cannot provide nitrogen.

12. (a) Mixing up of sand limestone, sodium carbonate, aluminium oxide and borax in required proportion.
(b) Addition of broken pieces of glass.
(c) Addition of ingredients, like, red lead, aluminium oxide manganese oxide, chromium oxide, etc.
(d) Melting mixture in a glass tank furnace or pot furnace.
(f) Blowing of glass
(e) Annealing process

13. Graphite is soft but this property is not exploited in the given applications.

14. RCC is very strong because of the presence of structural steel inside it. Hence, it is most suitable for the construction of pillars.

15. Fuse wire is made up of an alloy constituted of lead and tin in 1 : 1 ratio by mass as both lead and tin have comparatively low melting points than other metals.

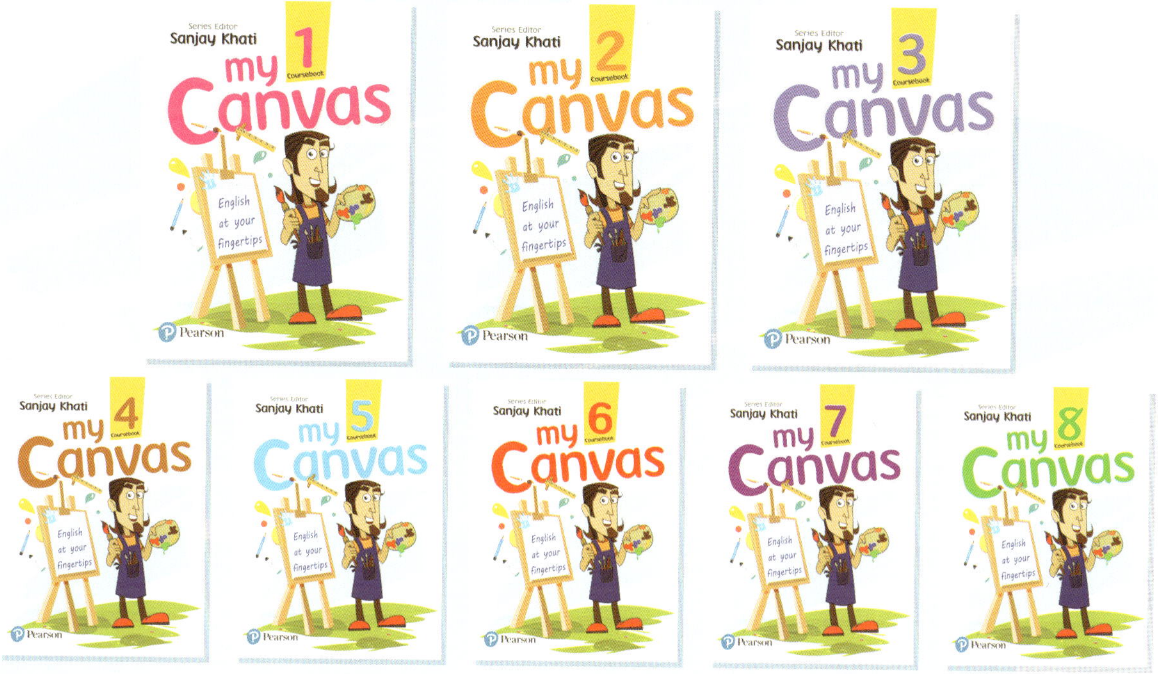

My Canvas is an all-inclusive course in English with a focus on developing skills and building language competencies. A blended learning tool, the course has been designed to teach language for effective communication. The focus on linguistic form and function provides holistic language learning, laying a firm foundation for using English confidently.

The coursebook contains fascinating, contemporary pieces of prose, poetry, essays, articles and drama to enrich the reading experience of young learners. In keeping with the blended approach, for the integrated development of the learner, each book has been divided into many classic and contemporary themes. The pre-theme pages help the learner connect with Multi-skill and MI activities. The Brush Up page reinforces and extends learning with reflection and life skills.

Different aspects of language learning are covered through these features in the coursebook.

- **Comprehension** promotes and assesses literal, inferential, predictive, analytical and evaluative understanding.
- **New Words** helps the learners retain unfamiliar words, phrases and their meaning.
- **Vocabulary** draws the learner's attention to the lexical and semantic wealth of language.
- **Grammar** aids the learners in understanding and applying the nuances of language, grammar and usage through graded and engaging explanations and exercises.
- **Pronunciation** facilitates the learners to master correct pronunciation and intonation.
- **Listening and Speaking** develops oral and aural skills and helps build confidence.
- **Composition** encourages the learner to master writing skills through guided and free writing tasks.

FREE SCHOOL SUPPORT MATERIALS

This book is supplemented by a Teacher's Resource Book and a Teacher's CD-ROM which contain

- E-book with interactivities and audio recordings of the poems, stories, plays, pronunciation drills and listening inputs
- **Animations** and **Worksheets** for use in the classroom
- **Lesson plans** with teaching guidelines for all the lessons
- A **Dynamic Question Bank** that allows teachers to generate question papers for various assessment purposes

For further information, please contact us at **schoolmarketing@pearson.com**.

Tune in to Grammar presents a systematic and step-wise strategy for approaching grammar at any grade. The series presents grammar rules lucidly and leads the learner to assimilate and apply them to communication in everyday life.

The learner is introduced to the concept with a warm-up and led through clearly demarcated stages of acquiring proficiency in grammar. The culmination of a chapter corresponds to the learner being able to use the language in games, projects, activities and a wide range of situations which involve working individually or in groups.

FREE SCHOOL SUPPORT MATERIALS

This book is supplemented by a **Teacher's Resource Book** and a **Teacher's CD-ROM** which contain

- answer keys
- grammar charts
- Teacher's CD with worksheets, animations and grammar games
- step-wise guidelines on how to conduct the Activity in the classroom

For further information, please contact us at **schoolmarketing@pearson.com**.

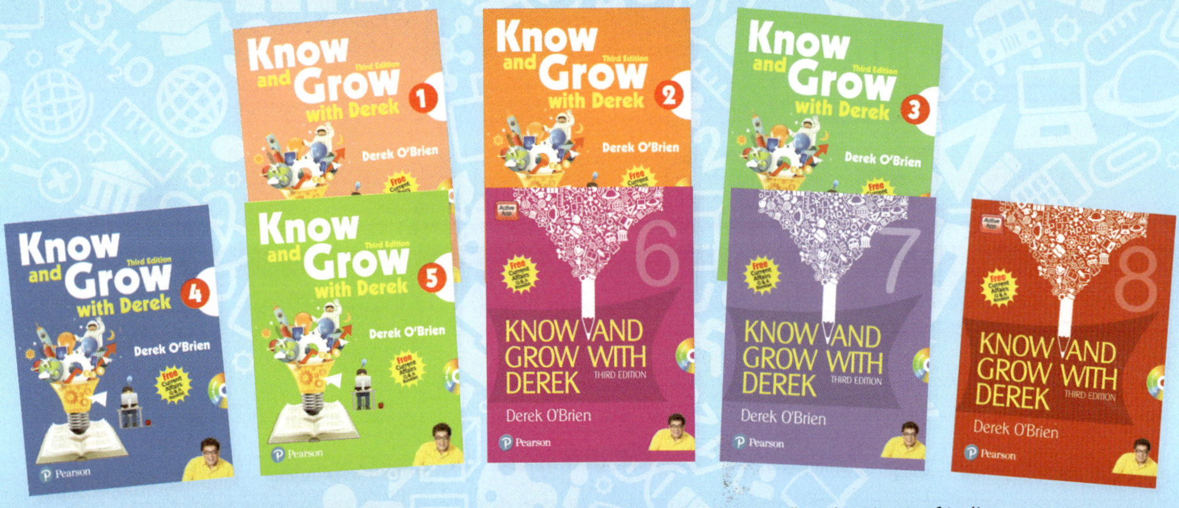

"In this country Derek O'Brien is synonymous with knowledge based game shows."—*The Times of India*

"There is a bright child in every child and Derek knows exactly how to bring it out."—*Oman Daily*

"Derek O'Brien's new mantra is *Know and Grow*... emphasising on the overall development, life skills and creativity of the child, along with general knowledge." —*The Arab Times, Kuwait*

"Derek O'Brien's way with children is inimitable." —*The Telegraph*

"A great believer in learning-by-doing, Derek O'Brien had set his eyes on general awareness to help students gather information in an interesting way." —*Indian Express*

"Derek O'Brien fills the mind-game with a dash of excitement and fun." —*The Hindu*

Know and Grow with Derek, 3e, is a series aimed at addressing the four vital skills for success in student life—Personality Development, Life Skills, General Knowledge along with Creativity and Thinking Skills. The series can be used as a self-help guide or can be teacher-driven. All useful, everyday information has been classified as KNOW and how to use information to get ahead in life as GROW. The *Know with Derek* and *Grow with Derek* boxes are used throughout the series to highlight different areas of information and growth. The format and design are child-friendly and easily accessible.

KEY FEATURES

- The topics are thematically divided into Units.
- The layout is attractive and student-friendly.
- Colourful pictures and illustrations supplement the textual content effectively.
- The *Know with Derek* section provides interesting trivia on a particular topic.
- The *Grow with Derek* section encourages emotional growth and creativity.
- Quizzes reinforce the learning of important concepts.
- Activity pages engage the learners in stimulating activities.
- Life Skill and Value Education section enhances the positive behaviour in a learner.
- Thinking and Number Skills section enhances the logical and mathematical acumen of learners.
- Fact-based chapters provide detailed information about selected topics.
- Test Papers evaluate the performance of the learners.
- Answer Keys are provided at the end of the book.

FREE SCHOOL SUPPORT MATERIALS

This book is supplemented by a **Teacher's Resource Book** and a **Teacher's CD-ROM** which contain

- **Dynamic Question Bank** enabling teachers to auto-generate question papers, are also provided.
- **E-book:** The e-book has been designed as a dynamic and interactive tool that will make learning more enjoyable. Its various features, such as zoom, highlight and search, aid classroom teaching and enhance understanding of concepts.

For further information, please contact us at **schoolmarketing@pearson.com**.